DYNAMICAL PROCESSES IN ATOMIC AND MOLECULAR PHYSICS

Editors

Gennadi Ogurtsov
A.F.Ioffe Physico-Technical Institute
Russia

Danielle Dowek
Université Paris-Sud
France

CONTENTS

FOREWORD

Atomic and molecular physics is now hundred years old, but contrary to peoples of the same age (even nowadays), this field remains surprisingly young. However, its life as that of humans suffered several ups and downs. After an exploding youth during its first 30 years, during which everything was thought to be discovered and understood, AMP fell into a winter sleep of about 20 years, when most of the scientists turned towards its younger sister, the nuclear physics. It is after the Second World War that AMP started a new life, a first rebirth that primarily involves Russia and USA physicists with the dream, still a dream nowadays, of a possible control of nuclear fusion in tokamak devices. A better understanding of the interaction between atomic species at high temperature was necessary that triggered a lot of research on atomic collisions. But, as in any field of research, breakthroughs come only with the availability of new tools. AMP did not made exception, and the new real take-off of the field came few years later with the discover of the laser, and since that time their continuous improvements lead, with each of them, to multiple new rebirths of AMD.

The eBook edited by G. Ogurtsov and D. Dowek provides several papers that review recent progress in the understanding of AMP related to the spectacular improvement in the exploration time of the basic atomic mechanisms. Pump probe spectroscopy in the femto- and atto-second regimes allows more and more accurate measurements of the localisation and motion of the active electrons, with a special interest in the most basic H_2 and D_2 molecules. New laser technologies also allow studying atomic interactions at extremely low temperature at which the quantum aspect leads to surprising atomic behavior. This was recently extended to the extreme cooling of molecules, formed by association of cold atoms.

Another tool is provided by the spectacular development of new imaging technologies with sophisticated coincidence particle detection devices that allow real advances in the understanding of photofragmentation and photoionization of small molecules after being fired by laser and synchrotron light sources. A spectacular development of many particle imaging technologies was provided by the invention of the “reaction microscope”. Its recent use in connection with intense V-UV femtosecond pump probe experiments allows time resolved imaging of molecule nuclear motions.

More generally, our understanding of the electron dynamics made spectacular progresses. New theoretical tools were invented with the introduction of complex wave functions in the time dependent Schrödinger equation showing the role of vortices in dynamical processes. The discovery of Coulombic decay involving interactions between two close atoms and molecules leads to large progress in the understanding of decay processes involving excited and ionized clusters.

Michel Barat

Institut des Siences Moléculaires
Université Paris-Sud
France

PREFACE

This eBook is aimed at highlighting the present state of experimental and theoretical studies in the field of atomic and molecular physics. Atomic and molecular physics provides a basis for our understanding of fundamental processes in nature and technology, which is also of interest for applications in other domains, such as solid state physics, chemistry and biology.

In recent years, atomic and molecular physics has undergone a revolutionary change due to tremendous achievements in computing and experimental technique. Now it is possible to study the processes inaccessible before and, in particular, correlated quantum dynamics, *i.e.,* time evolution of a quantum system composed of interacting atoms and molecules. The use of super-computers and novel experimental methods based on laser cooling, femto- and attosecond imaging technique have led to new discoveries in the dynamics of atomic and molecular processes.

The eBook includes four chapters, chosen to cover advanced directions in the modern atomic and molecular physics. Chapter 1 is devoted to the discussion of newly discovered phenomenon of vortices in electronic wave functions that are solutions of the time-dependent Schrödinger equation. Chapter 2 discusses the interatomic electron transitions in molecular systems, the sources of super-fast decay of these systems. Chapter 3 considers current perspectives on the study of molecular frame photoemission in one- and multiphoton photoionization of small polyatomic molecules. Chapter 4 introduces the reader to the recent achievements in laser cooling of atomic and molecular systems.

The primary target audiences of this eBook are those who are interested in atomic and molecular physics. They include researchers, developers and graduate students. The eBook could be a useful reference in university courses in atomic and molecular physics.

We would like to express our sincere appreciation to all the authors of the chapters of the eBook. We would like to thank Prof. Michel Barat for writing foreword, and the Bentham Science Publishers, in particular, Manager Bushra Siddiqui, for their support and efforts.

Gennadi Ogurtsov
A.F.Ioffe Physico-Technical Institute
Russia

Danielle Dowek
Université Paris-Sud
France

List of Contributors

Averbukh, Vitali

Department of Physics, Imperial College London, Prince Consort Road London SW7 2AZ, United Kingdom

Cederbaum, Lorenz

Physikalisch-Chemisches Institut, Universität Heidelberg, Im Neuenheimer Feld 229 D-69120 Heidelberg, Germany

Chiang, Ying-Chin

Physikalisch-Chemisches Institut, Universität Heidelberg, Im Neuenheimer Feld 229 D-69120 Heidelberg, Germany

Colorenč, Přemysl

Institute of Theoretical Physics, Charles University in Prague, V Holešovičkách 2, 180 Prague, Chech Republic

Comparat, Daniel

Laboratoire Aime Cotton CNRS, Université Paris-Sud, Bld 505, 91405 Orsay Cedex, France

Demekhin, Philipp V.

Institut fur Physik, Experimental Physik IV, Universitat Kessel, Heinrich-Plett Str. 40, D-34132 Kessel, Germany

Dowek, Danielle

Institut des Siences Moléculaires d'Orsay, Federation Lumiere Matiere, Bat.350, Université Paris-Sud, 91405 Orsay, France

Fioretti, Andrea

Laboratoire Aime Cotton CNRS, Université Paris-Sud, Bld 505, 91405 Orsay Cedex, France

Gokhberg, Kirill

Physikalisch-Chemisches Institut, Universität Heidelberg, Im Neuenheimer Feld 229 D-69120 Heidelberg, Germany

Kopelke, Sören

Physikalisch-Chemisches Institut, Universität Heidelberg, Im Neuenheimer Feld 229 D-69120 Heidelberg, Germany

Kryzhevoi, Nikolai V.

Physikalisch-Chemisches Institut, Universität Heidelberg, Im Neuenheimer Feld 229 D-69120 Heidelberg, Germany

Kuleff, Alexander I.

Physikalisch-Chemisches Institut, Universität Heidelberg, Im Neuenheimer Feld 229 D-69120 Heidelberg Germany

Lucchese, Robert R.

Department of Chemistry, Texas A&M University, College Station, Texas 77843-3255, USA

Macek, Joseph H.

Departmebt of Physics and Astronomy, University of Tennessee and Oak Ridge National Laboratory, TN 37996-1501 Knoxville, USA

Pillet, Pierre

Laboratoire Aime Cotton CNRS, Université Paris-Sud, Bld 505, 91405 Orsay Cedex, France

Scheit, Simona

Department of Basic Science, Graduate School of Arts and Sciences, The University of Tokyo 153-8902 Tokyo, Japan

Sisourat, Nicolas

Physikalisch-Chemisches Institut, Universität Heidelberg, Im Neuenheimer Feld 229 D-69120 Heidelberg, Germany

Stoychev, Spas D.

Physikalisch-Chemisches Institut, Universität Heidelberg, Im Neuenheimer Feld 229 D-69120 Heidelberg, Germany

CHAPTER 1

Vortices in Atomic Processes

Joseph H. Macek*

Department of Physics and Astronomy, University of Tennessee, Knoxville TN 37996-1501 and Oak Ridge National Laboratory, Oak Ridge TN, USA

Abstract: The time-dependent Schrödinger equation describes dynamical processes of one-electron species in terms of complex wave functions. The functions are inherently complex; therefore zeros occur only when both the real and imaginary parts of the wave functions vanish. If this happens at isolated points rather than on nodal surfaces one can show that the zeros must correspond to vortices. An imaging theorem is given which shows how such vortices can be seen experimentally. Since the theorem requires time propagation from microscopic to macroscopic scales, a method is developed that does just the same. Examples of vortices that emerge in dynamical processes are given. The vortices that we find are linked to the hydrodynamic interpretation of Schrödinger's time-dependent equation.

I. INTRODUCTION

The early theory of the 1910-1930 decade was a notable advance in our understanding of the physical world in that it explained the stability of matter. This initiated a revolution in physical theory that still continues today. Fortunately, for most of the physics of daily life, the physical theory is fairly settled. It is only necessary to forgo a fixation on minute descriptions of classical orbits and concentrate on the stationary states of non-relativistic quantum mechanics to interpret the physical world around us. With the time-independent Schrödinger, and/or Dirac equations it is possible to compute, in some level of approximation, most of the static properties of atoms, molecules, liquids and other forms of condensed matter. Alternatively, advances in the understanding of dynamical processes have been less marked, despite the fact that Born [1] showed how to bring such processes within the reach of Schrödinger's formulation of the quantum theory and in the process discovered a key element of quantum theory, namely, the probability interpretation of Schrödinger's wave function.

With the probability interpretation it proved possible to extract many dynamical physical quantities, for example, scattering and transition amplitudes, using the time-independent theory. There is thus a long tradition of treating dynamical processes using the time-independent Schrödinger equation in the non-relativistic domain. Despite this tradition the concept of time is always lurking in the background. Even though one can compute transition matrix elements without reference to time, it is necessary to invoke that concept to identify transition rates. Once identified, time hardly plays any further role in the description of dynamical processes in quantum mechanics. In essence, the time-independent Schrödinger equation is more fundamental than the time-dependent equation. This predominance of the time-independent theory is also apparent in practical calculations. There are many more examples of exact solutions of the time-independent Schrödinger equation than there are for the time-dependent equation. One of the purposes of this chapter is to connect time-dependent theory with time-independent theory and to discuss some fundamental differences in the structure of free-particle wave functions in the two representations.

Time-dependent wave functions are inherently complex since the Schrödinger equation is complex. Real initial functions can evolve to complex functions at later times. As a consequence, two-dimensional slices of the function may have isolated zeros. In contrast, similar slices of real, time-independent wave functions have nodal lines. A central emphasis of this chapter is the vortex structure at isolated zeros.

Fig. **1** shows a plot of the wave function of a hydrogen atom in the ground state subjected to the time-dependent field of a passing proton [2, 3]. The green surface marks the region of space where the absolute

***Address correspondence to Joseph H. Macek:** Department of Physics and Astronomy, University of Tennessee and Oak Ridge, National Laboratory, TN 37996-1501 Knoxville, USA; Tel: (865) 974 0770; E-mail: jmacek@utk.edu

Gennadi Ogurtsov and Danielle Dowek (Eds)

Figure 1: Three-dimensional image of a line of zeros in a time-dependent wave function for atomic hydrogen during proton impact. The green tube surrounds a line of zeros *i. e.* a vortex line, and the gold sphere locates the main probability density of the 1s initial state. The green extremities are where the magnitude of the wave function is comparable to its value on the tube surrounding the vortex line.

value of the wave function is less than a specified small value. As that value shrinks to zero, the tornado-like structure shrinks to a vortex line. Any plane passing through the green line has an isolated zero where the line intersects the plane. In this chapter we will discuss how such vortices emerge in solutions of the time-dependent Schrödinger equation and how such structures may be observed experimentally.

To interpret the vortices a new perspective on quantum theory is useful. That there are vortices in atomic wave functions is nearly as old as the quantum theory itself. Indeed, the subject goes back to Dirac [4] and has been discussed occasionally ever since [5, 6]. The key new feature that emerges from our work is that vortices in atomic wave functions are *observable* in "complete experiments" [7], *i. e.* in experiments where the state of the observed system is specified to the fullest extent theoretically possible. The COLd-Target-Ion-Recoil-Momentum-Spectroscopy (COLTRIMS) [8] is a prime example of a technique that could observe the vortices we discuss here. To make this connection we have introduced an "imaging theorem" [9] that relates observed momentum distributions to coordinate space wave functions at large times.

To reach the requisite large times it is necessary to integrate the time-dependent Schrödinger equation as the system moves from microscopic to macroscopic dimensions. We discuss one method [10] to do this.

A feature that has given new emphasis owing to the discovery of vortices in time-dependent wave functions is the hydrodynamic interpretation of quantum mechanics. The hydrodynamic interpretation was first given by Madelung [11] less than a year after Schrödinger's [12] publication in 1926. The hydrodynamic interpretation has not had a notable influence on the development of the quantum theory [5, 13], but is useful for connecting vortices in single-particle wave functions with vortices in the more familiar domain of fluid mechanics.

The fundamental nature of the time-independent theory is taken as a starting point for our treatment of vortices in quantum mechanics. We connect it to the time-dependent theory by first defining time.

Time is defined following Einstein, namely, time is what a clock measures [14]. The "clock" in this case could be oscillations of atomic states, conventional timepieces so essential for our daily life, lunar and planetary orbits, or even the expansion of the universe [15]. An essential feature of clocks as discussed, for example, by Wigner [14] is that they are macroscopic. Indeed Wigner computed that a "clock" must have a weight of the order of a fraction of a gram. The clocks that we will appeal to are not so precisely defined, however, our starting point is the description by Mott [16] and amplified by later writers [17, 18] that time is a classical concept and that its introduction into quantum mechanics is *via* a semiclassical approximation valid in the limit that the clock is "macroscopic" in a sense to be made clear later. In this sense time is practically essential to make contact between the microscopic theory of atoms and molecules and the observation of events on the macroscopic scale of actual experiments.

II. TIME IN QUANTUM MECHANICS

Time is introduced into the quantum mechanics simply by incorporating a "clock" as a part of the physical system in addition to the parts that constitute our main interest. This can be done in a general way, demonstrating that the details of the clock make no difference in final result [17]. Alternatively, one can use the simplest possible "clock" to introduce time since the exact nature of the timepiece is of peripheral importance. To that end we will suppose that time is measured by a time of flight technique [16]. That is, we suppose that if the mass and energy, or velocity of a particle is known, then its position $\boldsymbol{R} = \boldsymbol{v}t$ defines the time t. For simplicity, the coordinates of the clock are just the particle's position $\varsigma = vt$ on the z-axis. If H is the Hamiltonian for the system without the clock then the Schrödinger equation with the clock is,

$$\left(-\frac{\hbar^2}{2M}\frac{\partial^2}{\partial\varsigma^2}+H\right)\Psi = E\Psi \tag{2.1}$$

where M is the mass of the "clock" particle. Setting $\Psi = \exp\left[iK\varsigma\right]\psi$ with $E = \dfrac{\hbar^2K^2}{2M}$ in Eq. (2.1) gives the equivalent equation.

$$\left(-\frac{\hbar^2}{2M}\frac{\partial^2}{\partial\varsigma^2} - i\hbar^2\frac{K}{M}\frac{\partial}{\partial\varsigma}+H\right)\psi = 0 \tag{2.2}$$

Eq. (2.2) is fully equivalent to Eq. (2.1) but has a rather different appearance owing to the absence of the energy E on the right hand side. This is compensated for by the presence of the first derivative term on the left hand side. Using that $\hbar K/M$ = v, setting $\varsigma = vt$, and taking the limit that $M \to \infty$ gives the time-dependent Schrödinger equation;

$$H\psi = i\hbar\frac{\partial\psi}{\partial t} \tag{2.3}$$

where the Hamiltonian H may or may not depend explicitly upon the time variable t. In either case, the time-dependent Schrödinger equation emerges from the time-independent equation when a macroscopic clock is explicitly introduced. Note that the limit $M \to \infty$ with v held constant is considered as a macroscopic limit in this construction.

The "clock" introduced above is one example of how time is introduced in the quantum theory, namely, a passing particle exerts a time-dependent field on a microscopic species. This is clearly adequate to treat non-radiation electric and magnetic fields. External radiation fields, as produced by laser or other sources of electromagnetic radiation are treated in a time-independent quantized field theory [19] at the fundamental level. Briggs and coworkers [17] have shown that even in this case a time-dependent theory can be derived by suitable choices of clocks. We therefore regard the time-dependent Schrödinger equation as an approximation to a fully time-independent theory for all external fields.

While explicit time-dependence is unnecessary at the fundamental level, it proves to be quite useful in actual practice. For example, one can readily represent pulsed fields that turn on and off when the time-dependence appears explicitly. This is difficult to do when the external radiation field is represented by the time-independent interaction of field oscillators. Of course, the radiation field can only be approximated by a time-dependent external field in the limit that spontaneous emission is neglected. In all of the applications considered here, such neglect will be assumed without explicit justification. Rather, our focus is on the time evolution of atomic states and not on the dynamics of the radiation field.

III. BASIS SET METHODS

In those cases where the Hamiltonian H is time-independent the Schrödinger equation Eq. (2.3) has solutions with a simple phase factor.

$$\psi(\{\mathrm{r}\},\, t) = \phi(\{\mathrm{r}\})\exp[-\mathrm{i}Et/\hbar] \tag{3.1}$$

so that we recover the time-independent Schrödinger equation without the "clock" degrees of freedom.

$$\mathrm{H}\phi(\{\mathrm{r}\}) = \mathrm{E}\phi(\{\mathrm{r}\}). \tag{3.2}$$

Here E is energy different from the essentially infinite value of $E = \lim_{M\to\infty} \hbar^2 Mv^2 \to \infty$, appearing in Eq. (2.1). Solutions of Eq. (2.3) include bound states ϕ_m and continuum states ϕ_ε. It will be assumed that the center of mass motion is factored out and the remaining particle coordinates $\{\boldsymbol{r}\}$ number $3N$ as for N independent particles. The set symbol $\{\boldsymbol{r}\}$ indicates that the coordinate includes the spin variable. Associated with each N particle is a reduced mass. For simplicity we will consider that these particles are all electrons or possibly nuclei with a given, possibly time-dependent, coordinates and that the spin degrees of freedom in H all refer to electron coordinates. Since E is fixed the solutions are eigenstates of the energy operator H. To articulate the general theory as simply as possible it is assumed that H describes a one-electron species, which could be an atom or an H^+-like molecular ion. In this case the set of coordinates $\{\boldsymbol{r}\}$ becomes just one spatial coordinate r. Where needed, generalizations to more than one electron will be indicated with a minimum of mathematical detail.

The fundamental constants $\hbar$ and the electron's reduced mass μ will be maintained for clarity of exposition. Numerical results, given in Sec. VI will, however, be given in atomic units where one sets $\hbar = c = m_e = 1$ where $m_e \approx \mu$ is the electron's mass.

When H is time-dependent, separable solutions are no longer appropriate and one must, in general, employ numerical methods even for relatively simple systems. It is convenient to consider two types of time dependence, namely, time dependence due to an external field that acts on the atomic species for a finite length of time, for example, turning on at $-t_0$ and turning off at t_0 as for a pulsed light field, and the time dependence due to passing ions acting on target atoms. These two situations differ somewhat since in the latter case electron transfer can occur and the external field due to the ion never vanishes. In the former case it does so the treatment of that physical situation is conceptually simpler. We consider:

$$H(t) = T + V_T(\boldsymbol{r}) + H_{\text{int}}(t) \tag{3.3}$$

where T is the one-electron kinetic energy operator,

$$T = \frac{\hbar^2}{2\mu}\nabla^2 \tag{3.4}$$

and the subscript T denotes "target" so that $V_T(\boldsymbol{r})$ is the target potential. The interaction term $H_{int}(t)$ represents the interaction of a field that is external to the atom but interacts, in first approximation, with the atomic electrons only. In this case a common method for solving the time dependent Schrödinger equation is to expand $\Psi(\boldsymbol{r}, t)$ in a complete orthonormal set of discrete basis states $u_n(\boldsymbol{r})$. At this point in the formal development we only assume that the basis is complete.

$$\psi_i(\boldsymbol{r},t) = \sum_n u_n(\boldsymbol{r})a_{ni}(t) \tag{3.5}$$

where the index i denotes that at an initial time t_i the wave function $\Psi_i(\boldsymbol{r}, t)$ is equal to that of the initial state of the electron. We will suppose that the initial state is the ground state $\phi_i(r,t) = \phi_i(R)\exp\lfloor -iEt/\hbar \rfloor$ of the target Hamiltonian $H_T = T + V_T$:

$$H_T\phi_i = E_i\phi_i. \tag{3.6}$$

In that case the time-dependent Schrödinger equation becomes:

$$\sum_{n'} H_{nn'}(t) = i\hbar \dot{a}_{ni}(t) \tag{3.7}$$

$$H_{nn'}(t) = \langle u_n | H(t) | u_{n'} \rangle \tag{3.8}$$

with initial conditions at time t_i that,

$$a_{ni}(t) = \langle u | \psi_i(t_i) \rangle \tag{3.9}$$

In other words, one simply employs the time-dependent Schrödinger equation in matrix form. As long as the set is complete and the sum in Eq.(3.5) converges the matrix form is equivalent to Eq. (2.3). Note that the target Hamiltonian and associated eigenstates play a role only in defining the initial state before the interaction is turned on. Eventually, after the field has turned off, they will be used again to project onto final states. Such projections are straightforward when the final state of interest is a bound state. It is also possible to carry out the projection when the final state is a continuum state; however, accurate use of that procedure requires precise knowledge of the final state, which is often problematical. In Sec. IV we will describe the "imaging theorem" that avoids this problem.

In practice the infinite sum is truncated to a finite number of terms. The series truncation basically replaces the exact Hamiltonian by a model Hamiltonian defined on a finite basis. This replaces H by a model Hamiltonian defined on a finite basis, which is then solved as a set of coupled first order differential equations. One difficulty of the simple direct solution is the highly oscillatory nature of the coefficients. This is usually treated by diagonalizing H_T to obtain approximations $w_n(\boldsymbol{r})$ to the set of physical states $\phi_n(\boldsymbol{r})$. The new set is also a set of discrete states that represent bound and continuum states, in some approximation. The discrete states with negative energies approximate bound states of H_T. The discrete states at positive energy are not energy eigenstates of H_T and are often designated as "pseudo-states".

The expansion coefficients for this model problem now contain exponential factors $\exp[-i\varepsilon_n t/\hbar]$ when the interaction is turned off. By incorporating these factors explicitly in the new coefficients $b_{ni}(t)$. $\exp[-ie_n t/h]$ the diagonal elements $H_{nn}(t)$ are modified so that the matrix elements of the target Hamiltonian removed from Eq. (3.7) now give the amplitudes of transitions between bound states directly.

It is impossible to list all of the basis states that have been used for numerical calculations, even for one-electron species. Usually, some sort of atomic eigenstate expansion is employed; however, for one-electron hydrogen-like species the Sturmian set $s_n(\mathrm{r})$ has been used. This set is sufficiently special that it will be described in detail.

Sturmian basis states are basis states that use the coefficient of the potential as an eigenvalue;

$$(T + \lambda_n(E) V_T) s_n(r) = E\, s_n(r) \tag{3.10}$$

where $\lambda_n(E)$ is the Sturmian eigenvalue. Notice that the Sturmian eigenvalue is a function of E. This means that for every E there is an infinite set of basis functions $s_n(E)$, but they are not physical eigenstates unless the eigenvalue $\lambda_n(E)$ equals unity for some n, although the parameter E can be chosen $E = E_j$ so that one of the $\lambda_n(E)$ equals unity. In actual practice n stands for a set of quantum numbers say n_j, l_j, m_j in a partial wave representation and values of E_j can be chosen so that the lower value of λ_j is unity for each choice of l and m. Techniques to employ these types of sets have been extensively developed for application to one-electron atomic species [20,22]

The matrix of coefficients $a_{ni}(t)$ for all possible values of initial states i is called the time evolution matrix and is denoted by $U_{nn'}(t, t_i)$. Associated with this matrix is the time evolution operator,

$$U(t,t_i) = \exp\left[-i\int_{t_i}^{t} H(t')dt'\right] \tag{3.11}$$

where H represents the model Hamiltonian, and the exponential is taken in the sense that it represents a product of infinitesimal steps in time;

$$U(t,t_i) = U(t,t-\Delta t)U(t-\Delta t, t-2\Delta t).U(t_i+\Delta t,t_i) \tag{3.12}$$

and the number of steps become infinite as Δt vanishes.

The expression (3.12) is taken to define the time evolution operator in the general case. It will be used in Sec.V to describe the Lattice-Time-Dependent- Schrödinger-Equation (LTDSE) method [10] that we use to compute time-dependent wave functions. With this method, one seeks to approximate the transition operator and all of its matrix elements including matrix elements for ionization, *i. e.* the entire $U(t_f,t_i)$-matrix with high accuracy for initial and final times, t_i and t_f.

A second type of time-dependence is that appropriate for an ion-atom collision where the electron moves in the field of two charged nuclei with masses that are about 10^3 larger than the electron mass. In that case we can employ the relative motion of the incoming projectile ion P as our time-of-flight clock. Then H just becomes:

$$H = T + V_T(r) + V_P(r - R(t)) + V_{PT}(R(t)) \tag{3.13}$$

where $\boldsymbol{R}(t)$ specifies a classical trajectory for relative motion of the nuclei. In most of the work discussed here $\boldsymbol{R}(t)$ is taken to follow a classical straight line trajectory $\boldsymbol{R}(t) = \boldsymbol{b} + \boldsymbol{v}t$ where $\boldsymbol{b}$ is the impact parameter and v is the relative velocity. Other choices for $\boldsymbol{R}(t)$ are sometimes used. For example, the physical situation sometimes indicates that the projectile is significantly deflected by the target nuclei, but not strongly deflected by the atomic electrons. Then it is often appropriate to include this deflection explicitly in the classical trajectory $\boldsymbol{R}(t)$. For simplicity only the straight-line trajectory is considered here. Other "trajectories" can be analyzed as in Ref. [17].

We solve the time-dependent Schrödinger equation with H as given by Eq. (3.12) with initial conditions that the electron is attached to the target T in a bound state i and that the projectile carries no electrons. One of the first steps usually introduced is to remove the potential $V_{PT}(R(t))$ by a phase transformation $\exp\left[-i\int^{t} V_{PT}(\boldsymbol{R}(t'))dt'\right]$ of the wave function. For notational simplicity we still call the transformed wave function $\psi_i(r,t)$ as in Eq. (2.3) since the exact meaning should be apparent from the context. If the projectile is an antiproton or otherwise does not bind electrons, then the procedures described above apply, since the interaction effectively turns off when t is sufficiently large. Thus if t_f is chosen so that $\boldsymbol{R}(t_f)$ is significantly larger than the mean radius of the highest energy pseudo-state $w_{n_{\max}}(\boldsymbol{r})$ then the projectile potential is weak where most of the electron probability density is located. Then the time-dependent interaction $V_P(\boldsymbol{r} - \boldsymbol{R}(t))$ has effectively turned off.

For the more common physical situations where the projectile can bind electrons, there is an additional complication not encountered in interactions with electromagnetic fields. This additional complication is electron capture to bound states of P. One can still employ a mathematically complete orthonormal basis set, but now the diagonalization of the model H_T does not include bound states of the projectile. This means that one must project the time-dependent wave function onto the projectile eigenstates to obtain amplitudes for electron capture into eigenstates. While this is possible, in principle, it is a procedure that inherently suffers from convergence difficulties and is now generally avoided. Rather one gives up the orthonormal basis set and expands in both target and projectile centered basis states. To do this, one must consider the solution of the Schrödinger equation in the frame of the projectile with the target potential turned off.

It is known that solutions, including plane wave solutions, of the Schrödinger equation are not invariant under the Galilean transformation $\boldsymbol{r}' = \boldsymbol{r} - \boldsymbol{R}(t)$, $t' = t$ in the usual sense. Because the time derivative taken holding $\boldsymbol{r}$ fixed differs from the time derivative holding $\boldsymbol{r}'$ fixed, eigenfunctions$_j$ $\varphi_j(\boldsymbol{r} - \boldsymbol{R}(t))$ in the target frame have an additional phase factor $\exp[i(\hbar\mu\boldsymbol{v} \bullet \boldsymbol{r} + 1/2\, \mu v^2 t/h)]$ called a Bates-McCarrol [18, 23] translation factor. This phase factor is not present in the projectile frame $\boldsymbol{r}' = \boldsymbol{r} - \boldsymbol{R}(t)$, $t' = t$. Since the basis set is not orthonormal the model Schrödinger equation Eq. (3.7) now includes a time-dependent overlap matrix $N(t)$ with elements $N_{nn'}(t)\langle n,t|n',t\rangle$. The coupled-state method now applies as before.

When the projectile velocity is less than the mean orbital velocity of the electron in the initial state $\sqrt{\langle|2T/\mu|\rangle}$ then an expansion in terms of *adiabatic* eigenstates is more appropriate than atomic states, *i. e.* a basis of eigenstate of $H(t)$ with t held fixed. It is also convenient to obtain the adiabatic states in a coordinate frame that rotates with the internuclear line. The transformation to a rotating frame is effected by the operator $\exp[-i\omega(t)L_y]$ where the y axis is perpendicular to the plane of $\boldsymbol{b}$ and $\boldsymbol{v}$, $L_y = -\hbar\mu\boldsymbol{R} \times \boldsymbol{v}$.

The appropriate translation factors for the adiabatic basis are difficult to identify owing to the fixed value of t in the definition of the basis set. This problem was essentially solved by Solov'ev [25] who introduced scaled coordinates $\boldsymbol{q} = \boldsymbol{r}/R(t)$, a scaled time $d\tau = dt/R(t)^2$ and a phase factor $\exp\left[iq^2\dot{R}(\tau)/\mu R^2(\tau)\right]$ where

$$\dot{R}(\tau) = \frac{dR(\tau)}{d\tau} \qquad \textbf{(3.14)}$$

This phase factor meets all of the requirements of a translation factor for large internuclear distances $R(\tau)$ but is now known as an explosion factor as will be discussed in the next section. There it will be seen that an explosion factor more closely matches the time evolution of continuum functions excited by projectile impact than do the Bates-McCarroll plane wave- type translation factors.

The Schrödinger equation in the new frame is explicitly;

$$\left[T + R(\tau)\left(V_T(q - \alpha\hat{R}(\tau)) + V_P(q - \beta\hat{R}(\tau))\right)\right]\psi(q,\tau) = i\hbar\frac{\partial\psi(q,\tau)}{\partial\tau} \qquad \textbf{(3.15)}$$

where $\alpha+\beta = 1$ and are chosen so that the origin of coordinates is taken at the center of mass of the two nuclei and $\alpha R(\tau)$ is the distance from that origin to the projectile. The kinetic energy operator is understood to be given by Eq. (3.4) where derivatives are with respect to the scaled variable q. The origin of coordinates is sometimes taken at the center of charge of the two nuclei, but for straight-line trajectories the exact position of the origin along the internuclear axis is not important.

A Sturmian basis set may be defined with the coefficient of the potential R in Eq. (3.15) as an eigenvalue. If $\Omega_n(R) = R^2E_n(R)$ are the fixed nucleus eigenvalues of the scaled Hamiltonian then the Sturmian eigenvalues $R_n(\Omega)$ are roots of the equation $\Omega(R_n) = \Omega$ for fixed Ω. These functions have only been used for approximate calculations but they lead [24] to the important hidden crossing theory of Solov'ev [25]. That theory will not be discussed here since the present emphasis is on exact numerical calculations. Even with this focus, the hidden crossing theory is uniquely valuable since it allows one to interpret experimental and calculated data in terms of intersections of energy surfaces $E_n(R)$ in the complex R plane. An example of that usage will be given in Section V.

With the Hamiltonian of Eq. (3.13) the transition amplitudes $a_{ji}(\boldsymbol{b})$ are functions of the two-component vector $\boldsymbol{b}$. This vector is unobservable and one cannot experimentally set a value of $\boldsymbol{b}$. Wilets and Wallace [26] show that the appropriate variable is the scattering angle Θ between the target and projectile in the final state. Amplitudes a_{ji} in the angle are related to amplitudes $a_{ji}(\boldsymbol{b})$ in the time-dependent impact parameter representation *via* a Fourier transform based on the definition of the "clock" [18]. The transformation is:

$$a_{fi}(\boldsymbol{K}_{\perp},K_{\parallel}) = (2\pi)^{-1}\int a_{fi}(\boldsymbol{b},K_{\parallel})\exp\left[-i\boldsymbol{K}_{\perp}\bullet\boldsymbol{b}\right]d^2b \tag{3.16}$$

where $\hbar\boldsymbol{K} = \mu_{TP}(\boldsymbol{v}_i - v_j\boldsymbol{b}) + 2(E_t - E_i)\hat{v}/\mu_{PT}v = \hbar\boldsymbol{K}_{\perp} + \hbar\boldsymbol{K}_{\parallel}$ is the momentum transfer when the final state is an excited state of the target. Here μ_{TP} is the target-projectile reduced mass, v_i is the initial relative velocity and $v_j\boldsymbol{b}$ is the perpendicular component of the final relative velocity. The quantity $\boldsymbol{K}$ is resolved into components parallel and perpendicular to the incident velocity. It takes a different form when f is an electron transfer state associated with the projectile since some momentum and energy are associated with moving frames. Note that $K_{\parallel} = 2(E_f - E_i)/\mu_{PT}v$. The transformation is discussed in more detail in Sec. VI.

IV. FREE-PARTICLE DYNAMICS

The time-dependent framework given in Sec. III is tailored to the treatment of bound states in time-dependent processes. To discuss the continuum states we first review the quantum theory of free particles.

Use has already been made of time-independent wave functions for free particles, namely, the plane waves $\varphi_k(\boldsymbol{r}) = (2\pi)^{-3/2}\exp(i\boldsymbol{k}\bullet\boldsymbol{r})$. Since this is an energy eigenstate the time dependence is just the phase factor $\exp[-iE_kt/\hbar]$, where $E_k = (\hbar k)^2/2\mu$ and μ is the particle mass. Of course this particular function is only one such solution. A general solution is the superpostion $\int a(\boldsymbol{k})\psi_k(\boldsymbol{r},t)d^3k$ where $\phi_k(\mathrm{r.t}) = \phi_k\exp(-iE_kt/\hbar)$ is normalized on the wave vector,

$$\int\psi^*_{k'}(\boldsymbol{r},t)\psi_k(r,t)d^3r = \delta(\boldsymbol{k}-\boldsymbol{k}') \tag{4.1}$$

Any wave function $\phi_i(\boldsymbol{r},t') = \langle\boldsymbol{r}|\phi_i(t')\rangle$ specified at time t' defines a wave function $\Psi(\boldsymbol{r},t)$ at later times t according to

$$\Psi(\boldsymbol{r},t) = \int d^3k\psi_k(r,t)\langle\psi_k(t')|\phi_i(t')\rangle \tag{4.2}$$

The amplitude a_{fi} that the system will be found is state $\phi_f(r.t)$ at time t is just the projection of $\Psi(r,t)$ onto $\phi_f(r.t)$,

$$a_{fi} = \int d^3k\langle\phi_f(t)|\psi_k(t)\rangle\langle\psi_k(t')|\phi(t')\rangle. \tag{4.3}$$

The probability that the system will be found in the state $\phi_f(r.t)$at time t is just the squared amplitude $|a_{fi}|^2$. The quantity

$$K(t,t') = \int|\psi_k(t)\rangle d^3k\langle\psi_k(t')| \tag{4.4}$$

in Eq. (4.3) has been called the free-particle propagator since it "propagates" the initial state vector $|\phi_i(t')\rangle$ to later times. It is also known as the time-dependent Green's function since it is the Fourier transform of the time-independent Green's function. It also satisfies the inhomogeneous equation,

$$\left(H - i\hbar\frac{\partial}{\partial t}\right)K(\boldsymbol{r},t;\boldsymbol{r}',t') = -i\delta^3(\boldsymbol{r}-\boldsymbol{r}')\delta(t-t') \tag{4.5}$$

with delta-function source terms, just as for the Green's functions of time-independent theory. The coordinate space representation of $K(t,t')$ in Eq. (4.4) is

$$\langle r|K(t,t')|r'\rangle = K(r,t;r',t') \equiv \left(\frac{\mu}{2\pi i\hbar(t-t')}\right)^{3/2} \exp\left(i\frac{\mu(r-r')^2}{2\hbar(t-t')}\right) \quad \textbf{(4.6)}$$

where it is understood that $t - t'$ is positive.

The coordinate space propagator can be understood by considering the wave function that results if $\psi_i(r,t')$ represents a particle localized at the origin at $t' = 0$. Then $\psi_i(r,t') = \delta^3(r')$ and we have

$$\Psi(r,t) = \int K(r,t;r,t)\delta^3(r)d^3r = \left(\frac{\mu}{2\pi i\hbar t}\right)^{3/2} \exp\left(i\frac{\mu r^2}{2\hbar t}\right). \quad \textbf{(4.7)}$$

Equation (4.7) shows that, if a particle is localized exactly at a point $\boldsymbol{r} = 0$ at some time t, then at later times, even infitesimally small later times, it is represented by an outgoing spherical wave extending everywhere in space with an amplitude proportional to $t^{-3/2}$. The superscript 3 on the Dirac delta-function is shown explicitly to emphasize that it is a delta-function in three dimensions, *i. e.*

$$\delta^3(r) = \delta(x)\delta(y)\delta(z)\,. \quad \textbf{(4.8)}$$

It is apparent that a wave concentrated in a small region of space at some initial time immediately expands to fill all space at positive times. In a sense, the wave function explodes. Indeed the quantity in Eq. (4.7) is called an explosion factor and was employed by Solov'ev [25] as an appropriate replacement for translation factors employed when wave functions represent bound states referred to moving reference frames. With the explosion factor in Eq. (4.7), the emphasis is on how waves with positive energy propagate when they are initially confined to limited regions of space.

It is not apparent that the wave function in Eq. (4.7) preserves normalization. The lack of normalization occurs because $\delta^3(r)$ is itself too singular to normalize in the usual fashion, that is, according to

$$\int |\Psi(r,t)|^2\, d^3r = 1 \quad \textbf{(4.9)}$$

since $\int (\delta^3)^2\, d^3r$ does not exist.

To verify the correct normalization, it is necessary to consider a wave confined to a region of space that is of finite, but small, extent. For example, suppose that $\varphi_i\,(r.t') = u_i(x.t)v_i(y.t)w_i(z.t)$ where $u_i(x,t) = iN_x \exp[-\frac{1}{2}(x/x_0)^2]$, x_0. is small but not exactly zero, and $N_x = (\pi x_0^2)^{-1/4}$ is a normalization constant. Similar forms are used for the y and z components of the wave function. Then the wave function at later times $t>t'$ is

$$u_i(\boldsymbol{r},t) = \int K(x,t;x',t')u_i(x',t')dx' = \left[\frac{\pi}{x_0^2}\left(x_0^2 + i\frac{\hbar t}{\mu}\right)^2\right]^{-1/4} \exp\left[-\frac{x^2}{2(x_0^2 + i\frac{\hbar t}{\mu})}\right] \quad \textbf{(4.10)}$$

The x-dependent factor of the time-dependent probability is

$$P(x,t) = \left(\pi|x_0(t)|\right)^{-1/2} \exp\left[-2(x/|x_0(t)|^2\right] \quad \textbf{(4.11)}$$

where $x_0(t) = (x_0 + i\hbar t/\mu x_0)$. It is apparent that,

$$\int_{-\infty}^{\infty} P(x,t)dx = 1$$

so that normalization is preserved. This result also shows that after a time $t = x_0^2\mu/.\hbar.$ the wave packet has doubled in width.

A spherical Gaussian wave packet of width r_0 spreads out from the origin at a finite rate so that the normalization to unity is maintained at all times. The spreading rate is of the order of $\mu/2\hbar r_0$ so in the $r_0 \to 0$ delta-function limit the instantaneous "explosion" is recovered.

The Fourier transform of $\psi(\boldsymbol{r}.t)$, denoted here as $\tilde{\psi}(\boldsymbol{k},t)$, relates to the momentum distribution. Because $\tilde{\psi}(\boldsymbol{k},t)$, only changes in phase, the momentum space probability $P(\boldsymbol{k}.t)$ remains fixed at its value at the initial time indicating that the momentum distribution does not change as the outgoing wave propagates in free space. Thus we see that momentum distributions are extracted by taking the squared magnitude of the *momentum* space wave function after all fields have been turned off.

The momentum distribution can be related to the coordinate space wave function in another way. For very large times we expect quantum wave functions to reflect classical motion. In this case, classical motion is just given by $\boldsymbol{r} = \boldsymbol{v}t$. If we substitute $\boldsymbol{r} = \boldsymbol{v}t$ in the probability distribution,

$$\lim_{t\to\infty}\left[|\psi(\boldsymbol{r},t)|^2 d^3r\right]\Big|_{r=vt} = P(\boldsymbol{k})d^3k \quad \textbf{(4.12)}$$

where $\boldsymbol{k}=\mu\boldsymbol{v}/\hbar$ and take the limit as t becomes infinite we obtain

$$P(k) = |\tilde{\psi}(k,0)|^2 \quad \textbf{(4.13)}$$

This shows that the *coordinate* distribution at very large times is identical to the *momentum* distribution at a time $t = 0$ where the wave begins propagating in free space.

This feature of continuum wave functions can be illustrated by using an initial anisotropic Gaussian with widths in rectangular coordinates proportional to $x_0 \neq y_0 \neq z_0$. At initial times the momentum and coordinate distributions are inversely related since the widths in momentum space are proportional to $1/2x_0.1/2y_0,$. $1/2z_0,$. As time increases, the width in coordinate space changes according to Eq. (4.11) but eventually equals the widths in momentum space. Such evolution is shown in Fig. **2** where the squared magnitude of an anisotropic Gaussian in the plane $y = 0$ and $k_y = 0$ is shown at three different times. Since the dimension of the packet expands in coordinate space, we have plotted the distribution as a function of scaled coordinates $\boldsymbol{q} = \boldsymbol{r}/t$. Initially the anisotropy in coordinate space is opposite to that in momentum space. At later times the coordinate space anisotropy changes while the momentum space anisotropy retains its initial distribution. Eventually at very large times, the coordinate space distribution becomes identical with the initial (and final) momentum distribution.

V. THE IMAGING THEOREM

The limiting procedure provides a way to extract momentum distributions from time-dependent wave functions. The procedure can be formally demonstrated by substituting $r=vt$ into the time-dependent wave function using Eq. (4.6) for the propagator,

$$\psi(\boldsymbol{r} = \boldsymbol{v}t, t) = \int K(\boldsymbol{r} = \boldsymbol{v}t, t; \boldsymbol{r}', t')\psi(\boldsymbol{r}', t')d^3r' \quad \textbf{(5.1)}$$

and taking the limit as $t\to\infty$ to obtain,

$$\lim_{t\to\infty} t^{3/2}\exp\left[-iE_k t/\hbar\right]\psi(\boldsymbol{r},t) = \tilde{\psi}(\boldsymbol{k},t') \quad \textbf{(5.2)}$$

where

$$\tilde{\psi}(\boldsymbol{k},t) = \frac{1}{(2\pi)^3}\int \exp(-i\boldsymbol{k}\cdot\boldsymbol{r}')\psi(\boldsymbol{r}',t')d^3r' \qquad \textbf{(5.3)}$$

In many applications of time-dependent quantum mechanics the outgoing electrons move in time-independent external fields after a 'perturbation" is turned off, for example, in the fields of a residual ion after an electromagnetic pulse. Then the momentum distribution emerges only when t' in Eq. (4.12) is sufficiently large that the external field effectively vanishes. To employ smaller values of t' one must employ a propagator G appropriate for the target:

$$\begin{aligned} G(r,t;r',t') &= \sum_n^l \langle \boldsymbol{r}|n\rangle \exp[-iE_n(t-t')/\hbar]\langle n|\boldsymbol{r}'\rangle \\ &+\int \langle \boldsymbol{r}|\psi_{k'}^-\rangle \exp[-iE_k(t-t')/\hbar]\langle \psi_{k'}^-|\boldsymbol{r}'\rangle d^3k \end{aligned} \qquad \textbf{(5.4)}$$

When $\boldsymbol{r} = \boldsymbol{v}t$ is substituted into the propagator acting on $\psi(\boldsymbol{r},t')$ and the limit $t\to\infty$ taken it becomes apparent that the bound state terms vanish. The continuum part is evaluated in the stationary phase approximation, which becomes exact in the limit, to obtain,

$$\begin{aligned} &\lim_{t\to\infty} t^{3/2}\exp(iE_k t/\hbar)\int G(r=vt,t;r',t')\psi_i(r',t')d^3r' = \int \psi_k^{i*}(r',t')\psi_i(r',t')d^3r' = \\ &= \langle \psi_k^{(-)}(t')|\psi_i(t')\rangle. \end{aligned} \qquad \textbf{(5.5)}$$

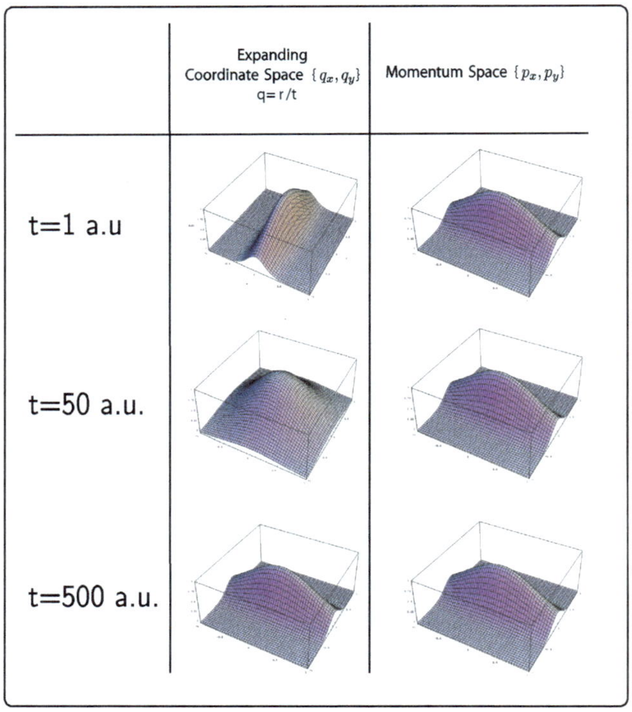

Figure 2: Coordinate and momentum space time-dependent wave functions comparing the evolution of anisotropic Gaussian packets in scaled coordinate space and unscaled momentum space.

In this case the momentum distribution is obtained by the limiting procedure of Eq. (4.12) or, as in Eq. (5.2) by projecting onto continuum eigenstates $\psi_k^{-*}(r',t')$ at $t = t'$. Sometimes, however, the continuum eigenstates are not known. In that case one may obtain the momentum distribution by taking the limit of the wave function directly. Equation (4.12), however, does not yield the ionization amplitude, rather Eq. (5.2) does when the external field vanishes sufficiently rapidly as $r \to \infty$. If there are long range potentials $V(r,t)$ present, then a well-defined limit in Eq. (5.2) is obtained if $E_k t$ is replaced by $\int^{t} (E_k - V(vt',t'))\,dt'$.

In actual practice this added complication plays no role in extracting momentum distributions. It is only for purposes of relating momentum distributions to wave functions that one must consider phases due to long-range potentials. To extract momentum distributions it suffices to take the limit of large times. Later in this manuscript we will illustrate one way to take the infinite time limit for solutions obtained by numerical procedures such as the Lattice-Time-Dependent- Schrödinger Equation (LTDSE) method [10].

To extract momentum distributions it suffices to take the limit of large times. Later in this manuscript we will illustrate one way to take the infinite time limit for solutions obtained by numerical procedures such as the Lattice-Time-Dependent-Schrödinger Equation (LTDSE) method [10].

Although our discussion has been limited to one-electron species, it generalizes fairly readily to systems where $\boldsymbol{r}$ represents a vector in N-dimensional space as for a system of $N+1$ particles with the center-of-mass motion factored out in the usual way. For example, if there are two electrons with coordinates $\boldsymbol{r}_1$, $\boldsymbol{r}_2$. in the field of a central ion, then the two-particle wave functions have two different types of continua, one where only one electron escapes and the other is bound, and the double continuum where both electrons escape. The substitution $\boldsymbol{r}_1 = \boldsymbol{k}_1 t$, $r_2 = \boldsymbol{k}_2 t$ followed by the $t \to \infty$ limit gives the double continuum momentum distribution $P(\boldsymbol{k}_1, \boldsymbol{k}_2) d^3k_1 d^3k_2$.

The one-electron continuum distributions are obtained by first projecting $\psi_i(r_1, r_2, t)$ onto the bound state wave function $\varphi_f(r_1,t)$ which is a time-dependent eigenstate of one electron in the field of the target ion. The $t \to \infty$ limit is taken on the projected function. In the above two-electron example it has been assumed that the two-electron spin function has been factored out and that the distributions correspond to specific spin states S where $\boldsymbol{S} = \boldsymbol{s}_1 + \boldsymbol{s}_2$. If spin is not a good quantum number or if more that two electrons are present, then it is necessary to include the spin labels in the wave functions and probabilities. This is a fairly standard task, so we will not illustrate it in this section. Finally, it should be noted that the discussion in this section applies to time-independent process, such as electron impact ionization, provided the initial state is represented by a wave packet localized in space. Then momentum distributions are obtained by taking the limit of wave packet that is sharp in momentum space after taking the $t \to \infty$ limit. We have seen that momentum distributions image the asymptotic coordinate space wave function according to Eq. (5.5). In this section that method will be related to momentum distributions obtained using standard, well-known, solutions of formal scattering theory.

To that end we write an exact time-dependent wave solution to the Schrödinger equation

$$\left(i\hbar\frac{\partial}{\partial t} - H\right)\Psi = 0 \qquad \textbf{(5.9)}$$

with incoming boundary conditions

$$\Psi_{u}^{(+)}(\boldsymbol{r},t) = \Phi_i^{(+)}(\boldsymbol{r},t) - i\int_{-\infty}^{\infty} G(\boldsymbol{r},t;\boldsymbol{r}',t')H_{\text{int}}(\boldsymbol{r}',t)\Phi_{u}^{(+)}(\boldsymbol{r}',t')d^3r'dt' \qquad \textbf{(5.10)}$$

where $\Phi_i^{+}(r,t)$ is an initial state solution for $H_0 = H - H_{int}$

$$\left(i\hbar\frac{\partial}{\partial t} - H_0\right)\Phi_i = 0 \qquad \textbf{(5.11)}$$

and H_{int} is an interaction Hamiltonian, and the propagator $G(\boldsymbol{r}, t; \boldsymbol{r}, t')$ is a Green's function satisfying

$$\left(i\hbar\frac{\partial}{\partial t}-H\right)G(\boldsymbol{r},t;\boldsymbol{r}',t')=i\delta^3(\boldsymbol{r}-\boldsymbol{r}')\delta(t-t') \tag{5.12}$$

with initial conditions $G(t,t')=0$, $t<t'$. Upon substituting $\boldsymbol{r}=\boldsymbol{v}t$ in the formal solutions Eq. (5.10) using the representation Eq. (5.4) of the Green's function and taking the large time limit we have

$$\lim_{t\to\infty}\Psi^{(+)}(\boldsymbol{r}=\boldsymbol{v}t,t)=-i\{(2\pi\}^{-3/2}\iint\exp\left[i(\boldsymbol{k}'\cdot\boldsymbol{v}t-E_{k'}/\hbar+\chi_k(\boldsymbol{v}t,t)\right]$$

$$\times d^3k'\left\langle\Psi_{\boldsymbol{k}}^{(-)*}(t')\middle|H_{\text{int}}(t')\middle|\Phi_i^{(+)}(t')\right\rangle dt' \tag{5.13}$$

where $\chi_k(r,t)$ is a phase factor to be discussed later.

The integral over d^3k' is evaluated by the stationary phase method which becomes exact as $t\longrightarrow\infty$. The stationary phase point is given by $\boldsymbol{k}'=\mu\boldsymbol{v}/\hbar = \boldsymbol{k}$ where it has been supposed that $\phi_{k'}(vt,t)/t$ vanishes in the limit of large time. This almost always holds, even for particles interacting *via* long range Coulomb interactions.

Introducing the transition amplitude

$$a_{ki}=-i\int\left\langle\Psi_{\boldsymbol{k}}^{(-)}(t')\middle|H_{\text{int}}(t')\middle|\Phi_i^{(+)}(t')\right\rangle dt' \tag{5.14}$$

into Eq. (5.13) and rearranging the phase factor gives the result

$$lim_{t\to\infty}\exp[iE_{\boldsymbol{k}}t/\hbar-i\chi_{\boldsymbol{k}}(vt,t)]\Psi^{(+)}(vt,t)=a_{\boldsymbol{k}i} \tag{5.15}$$

Eq. (5.15) shows that the transition amplitude "images" the time-dependent wave function for asymptotically large values of the electron coordinates $\boldsymbol{r}$ up to an overall time-dependent phase factor. For short range potentials the phase factor is just $exp[iE_kt/h]$ but for Coulomb potentials it is a logarithmic function that derives from the Redmond [28] phase factors. In any case, it does not affect the electron momentum distribution, which is the measurable quantity.

The imaging theorem is easily generalized to many particles. For example with two particles one has

$$\lim_{t\to\infty}[\psi(\boldsymbol{r}_1,\boldsymbol{r}_2,t)]^2\,d^3r_1d^3r_2\Big|_{r_1=v_1t,r_2=v_2t}=P(\boldsymbol{k}_1,\boldsymbol{k}_2)d^3k_1d^3k_2 \tag{5.16}$$

This equation gives the momentum distribution for double ionization. For single ionization with one electron in a bound state $\phi_m(\boldsymbol{r}_1)$ it is necessary to first project the full wave function onto the bound state. Calling this projection the "reduced" wave function $\psi_{rd,n}(\boldsymbol{r}_2,t)$ we have

$$\lim_{t\to\infty}\left[\left|\psi_{rd,n}(\boldsymbol{r}_2)\right|^2\right]\Big|_{r_2=v_2t}=P_n(\boldsymbol{k}_2)d^3k_2 \tag{5.17}$$

where $P_n(\boldsymbol{k}_2)$ is the electron momentum distribution corresponding to the one-electron final ion left in the n'th bound state. These equations are trivially generalized to any number of ejected electrons. Similar results hold for other types of fragments but are not given here.

The imaging theorem is very general and can be used to extract momentum distributions from numerically computed time dependent wave functions without employing an analytic asymptotic form for the wave functions. It can also be used to interpret momentum distributions in terms of time-dependent wave functions. This latter possibility is useful since some distributions may show unusual structure that can be best interpreted in terms of structures in coordinate space wave functions. We shall see that vortices may appear in time-dependent wave functions and these vortices give rise to unusual minima in electron momentum distributions.

VI. APPLICATIONS

To use the imaging theorem it is necessary to evaluate the wave function for very large times. This is not normally possible with typical time-dependent methods for solving the Schrödinger equation since such methods are designed to obtain accurate solutions in those regions where non-trivial dynamics occur. This is usually not at asymptotically large distances. Such tasks are usually handled by employing closed form analytic wave functions, e.g. projecting onto plane waves. Projecting onto plane waves usually requires first projecting out bound states since they have non-vanishing Fourier transforms that do not reflect the spectrum of ionized electrons. A better procedure is to project onto asymptotic eigenstates since the bound and continuum states are orthogonal. This can be done for simple time-dependent situations where the solutions are known or are easily obtained, however for particle transfer processes the selection of continuum wave functions that are orthogonal to all bound states is problematical [27].The alternative is to integrate the Schrödinger equation to infinite times.

A. Numerical Solutions at Large Times

In mathematically rigorous procedures the point at $t \to \infty$ is approached by defining a new variable $\tau = 1/t$ and going to the point $\tau = 0$. Use of the τ variable gives a version of the Schrödinger equation that simply moves the singularity at infinite time to another location. Fundamental obstacles still remain; in particular the divergent "explosion" factor is still present. This factor can be removed from the wave function and the solution, thus reduced, no longer oscillates rapidly at large *r*. The function, however, still expands to fill a large volume therefore to preserve normalization its magnitude must decrease accordingly. This means, for example, that at distances of the order of 1000 au the magnitude of the wave function is of the order of 10^{-9} compared with starting values of the order of unity, thus making it difficult to integrate to distances where Eq. (5.5) applies.

To circumvent the dimension problem one may scale the coordinates so that space expands with time. This does not lead to new singularities if the explosion factor has been removed. The corresponding theory closely follows the earlier hidden crossing theory of Solov'ev mentioned in Sec. IV. To make the transformations somewhat more general, scaling by a factor $R_s = \sqrt{b_s^2 + v_s^2 t^2}$ rather than the physical internuclear distance $R = \sqrt{b^2 + v^2 t^2}$ is employed. This has the advantage that scale of the coordinates in the region near $t = 0$ is selected by the parameter b_s, which can be chosen to obtain optimal precision in that crucial region while still maintaining the linear scaling with *t* for large times. The parameter v_s is included only for dimensional consistency since changing v_s is equivalent to changing the mesh of the time step Δt.

The three transformations made to obtain the equations of the Regularized Lattice Time Dependent Schrödinger Equation (RLTDSE) method are (1) transform to scaled coordinates $\boldsymbol{q} = \boldsymbol{r}/R_s(t)$ and (2) scaled times $d\tau = dr/R_s^2(t)$ then (3) remove the explosion factor to obtain the representation

$$\Psi(\boldsymbol{r},t) = R_s(\tau)^{-3/2} \exp\left[\frac{i\mu}{2\hbar R_s(\tau)^2} q^2\right] \Phi(\boldsymbol{q},\tau) \quad \textbf{(6.1)}$$

here $\Phi(\boldsymbol{q},\tau)$satisfies the transformed Schrödinger equabtion

$$[\frac{\hbar^2}{2\mu}\nabla_{\ddot{u}}^2 + R_s^2 V_s(\boldsymbol{q},\tau) + \frac{1}{2} v^2 b_s^2 q^2]\Phi(\boldsymbol{q},\tau) = i\hbar \frac{\partial \Phi(\boldsymbol{q},\tau)}{\partial \tau} \quad \textbf{(6.2)}$$

and $V_s(\boldsymbol{q},\tau) = V(\boldsymbol{q}R_s, t(\tau))$ is just the potential written in terms of the scaled coordinates and scaled time. The transformed equation has the same structure as the unscaled Schrödinger equation with an additional harmonic oscillator potential. This potential is attractive for real b_s, but since b_s could be any complex number the additional potential could be negative. In the applications reported here b_s is always real. The modified Schrödinger equation Eq. (6.2) is called the regularized Schrödinger equation.

The regularized Schrödinger equation is most efficient for treating the continuum part of $\Psi(t)$ and is not so well-suited to bound state components. To combine both the advantages of the LTDSE and the RLTDSE methods one may use the LTDSE equations to obtain $\Psi_\alpha^{(+)}(t)$ for $t < 0$ where α is an initial channel. A set of such functions for different channels β is obtained using the LTDSE method then $\Psi_\beta^{(-)}$ functions with incoming boundary are obtained using time-reversal and the functions $\Psi_\beta^{(+)}$. These functions are then used to find a solution at $t = 0$ which will evolve into a function with only continuum components.

The propagation of such a continuum function to large times is well-adapted to the RLTDSE method. A function at $t = 0$ that will evolve into a solution with only continuum components is given by

$$\Psi_\alpha(\boldsymbol{r},0) = \Psi_\alpha^{(+)}(\boldsymbol{r},0) - \sum_\beta \Psi_\beta^{(-)}(\boldsymbol{r},0) S_{\beta\alpha} \tag{6.3}$$

$$\text{where } S_{\beta\alpha} = \left\langle \Psi_\beta^{(-)}(t) \middle| \Psi_\alpha^{(+)}(t) \right\rangle \tag{6.4}$$

is an S-matrix element. This element is independent of time and will be calculated for $t = 0$ in applications.

This function defined at $t = 0$ is propagated forward in time numerically to obtain a continuum outgoing wave and the imaging theorem Eq. (5.15) is used to extract momentum distributions. This optimal use of the LTDSE method has given highly accurate results for all cross sections including momentum distributions for proton impact on atomic hydrogen for protons in the velocity range 0.2 a.u.< v < 2 a.u.

The propagation in time is performed using the LTDSE computer programs since both the original Schrödinger equation and the regularized version have the same form. It is only necessary to employ the τ-dependent interaction potential of Eq. (6.2) and propagate in the new time-like variable τ. This is accomplished as in the LTDSE with the variable τ, that is, propagation over one time step Δt is effected using the representation

$$\exp\left[-i\int_{t-\Delta t}^{t+\Delta t} H(t')dt'\right] \approx \text{txp}(-iT\Delta t/2)\left\{\exp\left[-i\int_{t-\Delta t}^{t+\Delta t} V(t')dt'\right]\right\}\exp(-iT\Delta t/2) \in \tag{6.5}$$

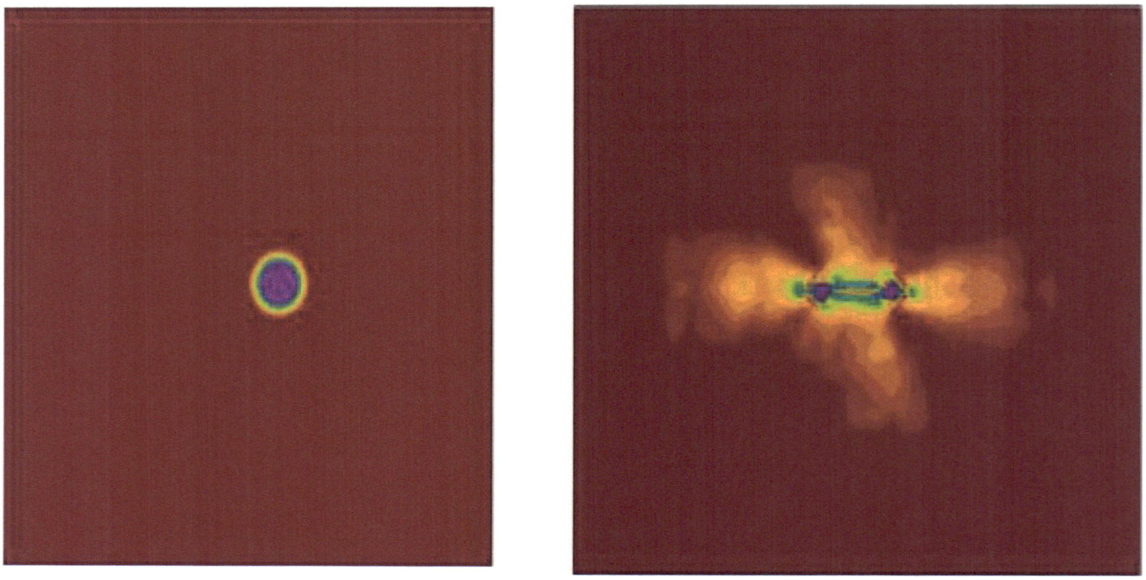

Figure 3: Contour plots in the y = 0 plane of the coordinate space wave function for proton impact on atomic hydrogen at initial times when $R \approx 10$ a.u. and at final times when $R = 5 \times 10^5$ a.u. Coordinates are scaled to the internuclear distance R(t). The box in the first panel is 20 a.u. on each side while the box in the second panel is 5×10^6 a. u. on each side.

where $T = -\frac{\hbar^2}{2\mu}\nabla^2$ is the electron's kinetic energy operator, to propagate the wave function from $t - \Delta t/2$ to $t + \Delta t/2$. The operator in Eq. (6.5) acts on the wave function $\psi(\boldsymbol{r}, t - \Delta t/2)$ to give the wave function at $t + \Delta t/2$. The iterated operator effectively involves the potential term times the kinetic energy term exp[$-iT\,\Delta$t] so that only the product of two operators are involved except for the first and last interactions. The LTDSE method uses a Fast-Fourier-Transform (FFT) algorithm to transform $\Delta(\boldsymbol{r}, t - \Delta t/2)$ to the momentum representation, then uses the diagonal form of $T = -\boldsymbol{p}^2/2\mu$ where $\boldsymbol{p} = \hbar\boldsymbol{k}$, followed by a back transform to coordinate space. Since the potential term is diagonal in coordinate space, its effect is to multiply the wave function by a phase factor giving a wave function at $t + \Delta t/2$. The process is repeated to step through time in intervals of Δt. Errors of the time step are of order of Δt. Errors of the implicit integrations over coordinates are not known precisely, however, the FFT and back FFT seem to cancel out errors to a greater extent than expected on purely formal grounds. In any case, the method is stable, flexible, and can be used with both the time-dependent Schrödinger equation and the regularized τ-dependent Schrödinger equation.

Of course, the high accuracy has a price, namely many time-consuming calculations of $S_{a}e$ are involved. The high price can be appropriate for benchmark purposes. Even then, it is only practical to subtract off the dominant bound state channels in Eq. (4.2). Indeed early applications of the LTDSE method to excitation and electron transfer processes gave satisfactory results. Only for ionization there were indications that the method had limitations for obtaining momentum distributions. If less accuracy is acceptable, as for semi-quantitative interpretation of electron momentum spectra, the RLTDSE method with appropriately chosen values for b_s can be used on the compete time interval from an initial time to to macroscopic final time intervals. This is particularly useful for examining the complete evolution in time of the wave function.

To demonstrate that the RLTDSE method allows one to propagate the electron wave functions from microscopic distances of the order of 10 a.u. to macroscopic distances of the order of 10^5 a.u. two slices of the time dependent wave function for proton impact on atomic hydrogen are shown in Fig. **3**. The left panel shows the $y = 0$ slice of the initial 1s atomic hydrogen wave function centered on the target nuclei. The distance scale is chosen so that the figure measures 20 x 20 a.u. At this point one just sees a contour plot of the spherically symmetric 1s wave function with the projectile nucleus separated from the incoming proton by about 10 a.u. Since only the electron wave function is plotted, the position of the incoming proton is not apparent. The rightmost panel shows a contour plot of $|\Phi(\boldsymbol{q}, t)|$ after the target and projectile nuclei have separated by a distance of 5×10^5 a.u. or about 26 μm, a macroscopic distance that is comparable to the dimensions of a human hair. Propagation in time to even greater distances is possible, but since there are essentially no changes in $|\Phi(\boldsymbol{q}, t)|$ after an internuclear separation of 3000 a.u. the distribution has reached it asymptotic value. It will be noted that Rydberg states with n of the order of 50 are confined to one pixel in the rightmost panel. At 3000 a.u. Rydberg states with n less than 5 occupy a space of about 100 pixels near the target and projectile nuclei. Distributions with those dimensions can change since time-dependent superpositions of bound states oscillate in time. Such oscillations are too small to be seen on the scale of the second panel in Fig. **3**. Even so, the figure illustrates the stability of the RLTDSE method.

B. Vortices in Momentum Distributions

In a first application of the regularized equation, *ab initio* calculations [3] verified the π-type molecular orbital component of the analytical hidden crossing theory [24] at 5keV impact energies where the adiabatic basis should be applicable. In this latter theory, electrons are promoted from bound state σ-molecular channels to π-bound states centered near the top of the potential barrier between incoming and target protons. As the species separate, the π distribution remains and gives the momentum distribution seen in Fig. **4**. The electron distribution in the hidden crossing theory also has a σ component and these two components interfere to give distributions seen in the *ab initio* calculations but which are difficult to model in the hidden crossing theory. Similar π-shaped momentum distributions have been seen experimentally for single ionization by proton impact on atomic helium. Indeed it was these experiments that stimulated the calculations of Refs. [3, 9, 10]. The momentum distribution shown in Fig. **4** is actually a plot of the magnitude $A_k(\boldsymbol{b}) = a_{ki}(\boldsymbol{b}, K_{||})$ of the coordinate space wave function at $b = 1$ when the target and projectile nuclei are separated by a distance of 25,000 a.u.

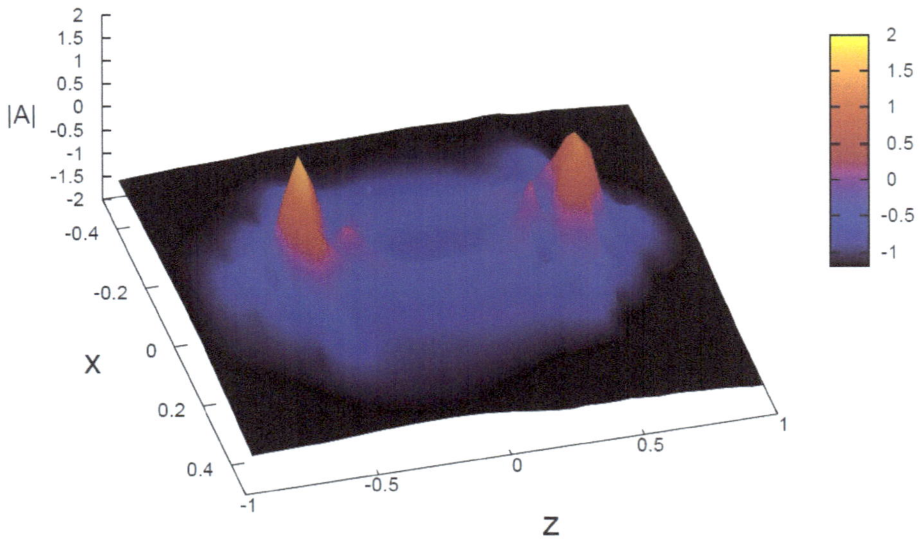

Figure 4: Momentum distribution produced by proton impact on atomic hydrogen at b = $1 a.u.$ and $E_P = 4keV$. The plot shows log $|A_k|$ and the color code of its magnitude is shown in the vertical strip on the right of the Figure The x and z axes give the x and z components of $\boldsymbol{k}/\mathrm{v}$ for the slice $k_y = 0$.

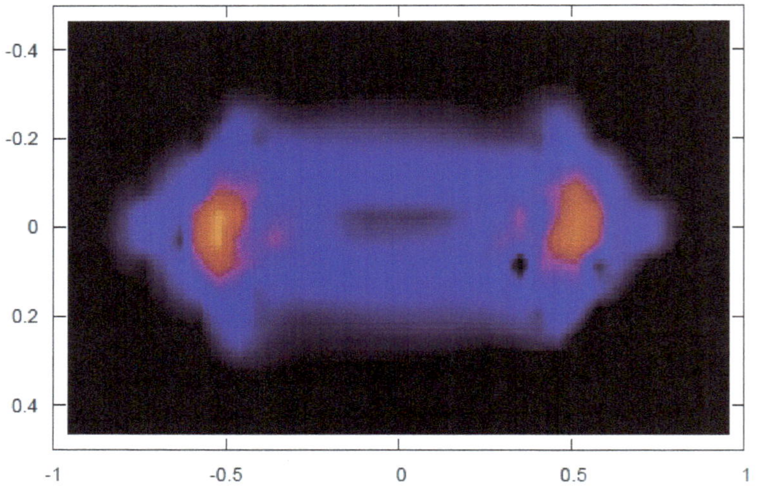

Figure 5: Contour plot of the electron distribution shown in Fig. **4**. The color scheme and coordinate labels are the same as in Fig. **4**. The "holes" show up more clearly in this plot.

One of the unexpected features of the momentum distribution shown in Fig. **4** is the presence of isolated "holes" where $P(\boldsymbol{k})$ vanishes exactly. To see these holes more clearly a contour plot of the distribution in Fig. **5** is shown. The origin of these "holes" has been traced to the formation of vortices in the time dependent wave function $\psi(r,t)$. Fig. **6** shows a contour plot of the wave function for proton impact on atomic hydrogen at 5 keV impact energy and an impact parameter of 1 a . u. Fig. **6a** shows the electron charge cloud at a distance of 5 a. u. between target and projectile nuclei where the electron probability is mainly near the target, but there is a non-negligible portion that has transferred to the projectile. At this point, the hidden crossing theory indicates that the $3d\sigma$ molecular orbital is excited as reflected in the shape resembling a Y_{20} $(\boldsymbol{r})$ function. The nodes of this function become more pronounced as the collision proceeds, Fig. **6b** where an isolated zero has split off from the node in the lower right hand corner of the Figure A blow-up of this quadrant is shown in Fig. **6b**' where arrows indicating the direction of the velocity current are shown. It is seen that the current flows around the isolated zero indicating that the zero does indeed correspond to a vortex. This vortex, thus formed, gives the tornado structure seen in the three dimensional image of Fig. **1**.

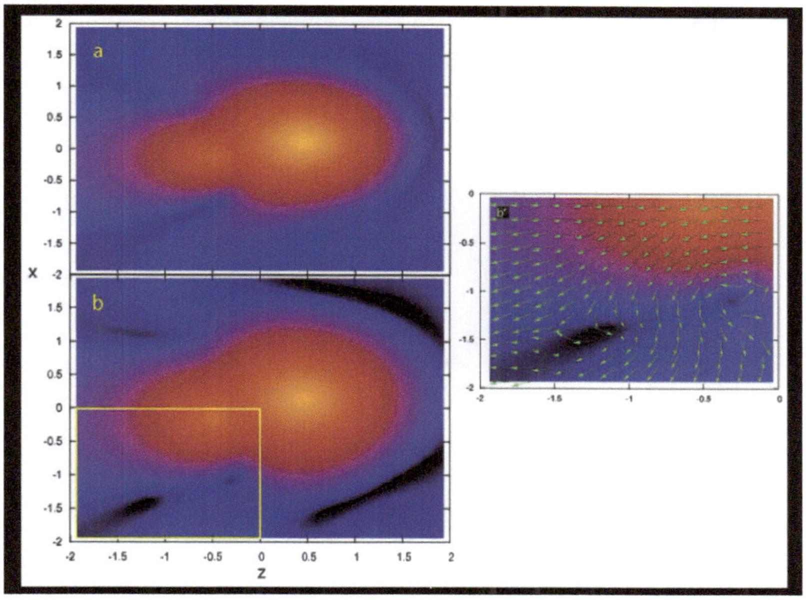

Figure 6: Vortex origin. (a) Density plot of the electronic wave function in scaled coordinates q at $R = 5$ a.u. All of the coordinates are scaled by the internuclear distance R with the coordinate z is parallel to the approaching protons initial velocity. The nucleus of the target (T) is centered at $q_z = 1/2$ and the projectile (P) is centered at $q_z = -1/2$. The bright areas indicate high electronic probability density and the dark areas low. Nodal lines are seen as the dark, linear features to the left of the projectile. (b) At $R = 4.8$ a.u. a hole, visible in the lower left-hand boxed quadrant, has separated from the nodal line. (b') Arrows represent the electronic probability current.

To understand how vortices appear in time-dependent Schrödinger wave functions it is useful to connect such structure with vortices in context of fluid flow, where they are more familiar. In the next section we review the hydrodynamic representation of the Schrödinger equation.

C. Hydrodynamic Representation of the Schrödinger Equation

Fluid flow is governed by the Navier-Stokes equations. For our purposes the most general form of these equations is not needed, rather a special form where both the viscosity and the bulk viscosity of the fluid vanish, are used. This special case is one of Euler's equations [29] for fluid flow for an ideal fluid with density ρ

$$\frac{\partial v}{\partial t} + \boldsymbol{v} \cdot \nabla \boldsymbol{v} + \frac{\nabla \boldsymbol{p}}{\rho} - \frac{\boldsymbol{f}}{\rho} = 0 \qquad \textbf{(6.6)}$$

where ***f*** is the force per unit volume.

Eq. (6.6) is essentially Newton's equation for the acceleration of a fluid element under the influence of a force due to the pressure p and force ***f*** per unit volume due to an external field.

Note that *f/p* has the dimensions of force per unit mass. In addition there is the equation of continuity expressing the conservation of mass:

$$\frac{\partial}{\partial t} + \nabla \cdot (\rho v) = 0 \qquad \textbf{(6.7)}$$

In the present case we will assume that the flow is irrotational except at isolated points [13]. Then the term v • ∇ v may be written

$$\boldsymbol{v} \bullet \nabla \boldsymbol{v} = \frac{\boldsymbol{1}}{\boldsymbol{2}} \nabla (v \bullet v) + (\nabla \times v) \times v = \frac{1}{2} \nabla (v \bullet v) \tag{6.8}$$

since the vorticity $\omega = \nabla \times \boldsymbol{v}$ vanishes for irrotational flow. In the present case it is assumed that ω vanishes everywhere except at isolated points where it may diverge. In that case the mean value of ω integrated over a small area d^2a encompassing these isolated points does not vanish.

Since the flow is irrotational one may set $v = C_1 \nabla S$ where C_1 is some constant, usually incorporated into the definition of *S*. To make connection with the Schrödinger equation, we will not follow this practice, but will instead set $C_1 = 1/\mu$ where μ has the dimensions of mass. Thus we define a quantity *S* according to

$$\boldsymbol{v} = \frac{1}{\mu} \nabla S \tag{6.9}$$

Also the term $\boldsymbol{f} / p$ which has dimensions of force per unit mass will be set equal to $-\nabla V / \mu$, where *V* is the, possibly time-dependent, potential of an external field.

$$\frac{\boldsymbol{f}}{p} = -\nabla V \frac{1}{\mu} \tag{6.10}$$

Equations (6.6) and (6.7) represent four equations with five unknown quantities ρ, v, and p so that one more equation is needed. This additional equation usually expresses some knowledge about the nature of the fluid. For example, if the fluid is incompressible then ρ is a constant. The pressure term is written as $\nabla\left(\frac{p}{\rho}\right)$ so that Euler's equation can be integrated directly to obtain the well-known Bernoulli's equation of incompressible fluid flow [29]. As is well known, the solutions of this equation describe streamline flow with no vortices.

A second type of fluid that is often considered is a compressible fluid where the density depends linearly upon the pressure $p=C_2\rho$. An ideal gas at constant temperature is an example of such fluid. In this case the pressure term can also be written as a gradient $\nabla p / p = C_2 \nabla(\ln \rho)$ and a first integral similar to Bernoulli's equation is obtained. This gives two equations for two scalar unknowns, ρ and *S*, which is a considerable simplification over the vector equations, although only approximate solutions are known in closed form. These solutions show wave motion, i. e. sound, but no vortex motion [29].

The final fluid we consider has no simple, classical realization. Rather a pressure and density relation is assumed that gives the time-dependent Schrödinger equation. It would be appropriate to say the Schrödinger equations describes a Schrödinger fluid or S-fluid. Such usage, however, suggests a real observable media, which does not exist. Rather we will call the resulting entity an S-field in keeping with current usage [30] where nature is described by quantized fields. Of course, we do not quantize the S-field since particle creation and annihilation are not involved in non-relativistic quantum mechanics.

The pressure-density relation that we assume, with the definition of the amplitude *A* as $p = A^2$, is

$$\frac{\nabla \mathrm{p}}{\rho} = \frac{\hbar^2}{2\mu} \nabla \left(\frac{\nabla^2 \mathrm{A}}{\mathrm{A}} \right) \tag{6.11}$$

Substituting Eqs. (6.11) and(6.7) into (6.6) gives

$$\nabla\left[\frac{\partial S}{\partial t}+\frac{1}{2\mu}(\nabla S)^2-\frac{\hbar^2}{2\mu}\frac{1}{A}\nabla^2 A+V(r,t)\right]=0 \quad \textbf{(6.12)}$$

while the equation of continuity becomes

$$\frac{\partial \mathrm{A}}{\partial \mathrm{t}}+\frac{1}{2\mathrm{m}}\mathrm{A}\nabla^2\mathrm{S}+\frac{1}{\mathrm{m}}\nabla\mathrm{S}\cdot\nabla\mathrm{A}=0 \quad \textbf{(6.13)}$$

Introducing the complex function

$$\psi = A\exp(iS/\hbar) \quad \textbf{(6.14)}$$

one sees that ψ satisfies the Schrödinger equation if ρ = A^2 and $\boldsymbol{v}=\nabla \boldsymbol{S}/\mu$ satisfy the Navier-Stokes equations for ideal irrotational field flow for a fluid with the S-field equation of state.

While the general S-field equation of state appears somewhat artificial, in the special case when $\rho=\exp(-\alpha r^2)$, as for the ground state of a harmonic oscillator, one can show from Eq. (6.11) that the pressure is given by $p=\rho/a$. In this case, the S-field appears much like an ideal gas although the Schrödinger theory requires a harmonic oscillator potential to keep the ideal gas relation valid for all time. It is somewhat remarkable that a simple pressure-density relation Eq. (6.11) gives a fluid-flow equation which has vortex solutions with no viscosity. When both $Re\psi(\boldsymbol{r},t)$ = 0 and $Im\psi(\boldsymbol{r},t)$ = 0 two different nodal surfaces are defined and the different nodal surfaces can intersect along a curve. Then two-dimensional slices through $|\psi(\boldsymbol{r})|$ may intersect the nodal line giving an isolated zero in the 2-d slice. At the isolated zero, the velocity field $\boldsymbol{v}=\nabla S/\mu = Im\nabla\ln\psi/\mu$ diverges and the curve is a vortex line. The integral of the vorticity $\omega=\nabla\times\boldsymbol{v}$ over any surface passing through the vortex line is found to equal 2π. That is

$$\int \omega\cdot\boldsymbol{n}d^2a=\oint \boldsymbol{v}\cdot d\boldsymbol{l}=\frac{2\pi\hbar}{\mu} \quad \textbf{(6.15)}$$

when ψ vanishes linearly at the zero. If ψ vanishes with the n'th power of $\boldsymbol{r}\rightarrow\boldsymbol{r}_0$ at $\boldsymbol{r}=\boldsymbol{r}_0$ then the integral equals $2\pi n$, thus the vortices of the Schrödinger equation are quantized vortices [6]. This is easily understood in the context of the solutions of the free-particle Schrödinger equation since, near any origin, there are solutions where angular momentum and the projection $n\hbar$ of angular momentum along an axis are good quantum numbers. These solutions vanish as r^n at the origin which can be any point $\boldsymbol{r}_0$. Thus there is a close association of the isolated zeros of ψ with angular momentum. The imaging theorem shows that these isolated zeros are observable and therefore provide information on the angular momentum properties of ψ.

If the potential is time-independent then the solutions of the Schrödinger equation can be taken to be real aside from a space-independent phase factor $exp(-iEt/\hbar)$. The resulting equation for A is then simply the eigenvalue equation for the energy eigenstates, which can be taken to be real. In that case the imaginary part of ψ vanished identically and the equations $\psi(\boldsymbol{r})=0$ defines a nodal surface. On any two-dimensional slice through the wave function there is a line of zeros, *i. e.* a nodal line, but there are no isolated zeros (vortices) on any two-dimensional slices of the wave function.

If we write $\Psi(\boldsymbol{r},t)=A(\boldsymbol{r},t)\exp[-iS(\boldsymbol{r},t)]$, where $A(\boldsymbol{r},t)$ and $S(\boldsymbol{r},t)$ are real, and it is found that $A(\boldsymbol{r},t)$ vanishes at an isolated point, that point must be on vortex line since then both $Re\psi$ and $Im\psi$ vanish and $\boldsymbol{v}=\nabla S$ has a pole. Therefore, isolated zeros in momentum distributions can be traced back to vortices in coordinate space wave functions using the imaging theorem. This gives a new classification [31] of zeros in electron momentum distributions, namely, nodal structures and vortices [32]. The former structures are well-known; however, studies of vortices in observable one-electron momentum distributions have only begun. The discussion has emphasized the evolution of vortices in calculations of time-dependent wave functions to give holes in ionization amplitudes a_{ki}. The process can be inverted, so that if a_{ki} is computed by standard means and holes are found, then these holes may be interpreted as coming from vortices in a time-dependent atomic wave function. The Born amplitude for ionization of the ground state of atomic hydrogen by proton impact is instructive in this regard.

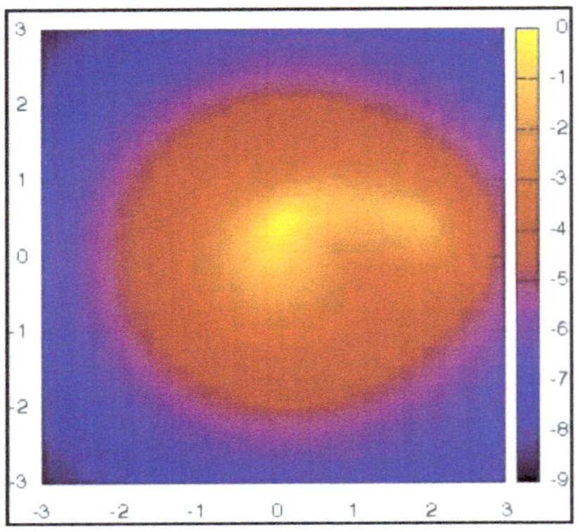

Figure 7: Contour plot of the electron momentum distribution computed in the plane-wave Born approximation (B1) for 400 keV proton impact on atomic hydrogen at a laboratory frame deflection angle of 5°. The numbers on the axes give the scaled electron momentum components k_x/v (vertical axes) and k_z/v (horizontal axis) in $k_y = 0$ plane. The vertical strip gives the color scheme for the log $|T_{ki}|$ plot.

The plane wave Born amplitude, T_{ki}^{B1} $(\boldsymbol{K})$, is given in standard texts [33, 34]. There are no points where it vanishes; therefore there are no vortices. Fig. **7** shows a contour plot of the $B1$ electron distribution in the x — z plane verifying the absence of vortices, in keeping with the fact that there is no net transfer of angular momentum between projectile and target in this approximation.

The $B1$ amplitude does not correspond to the amplitude calculated using the time- dependent Schrödinger equation Eq. (3.15) of Sec. III since the impact parameter b nowhere appears. The impact-parameter Born approximation $a_{ki}{}^{B1}(\boldsymbol{b},\boldsymbol{K}_{||})$ can be obtained by solving the first order equation.

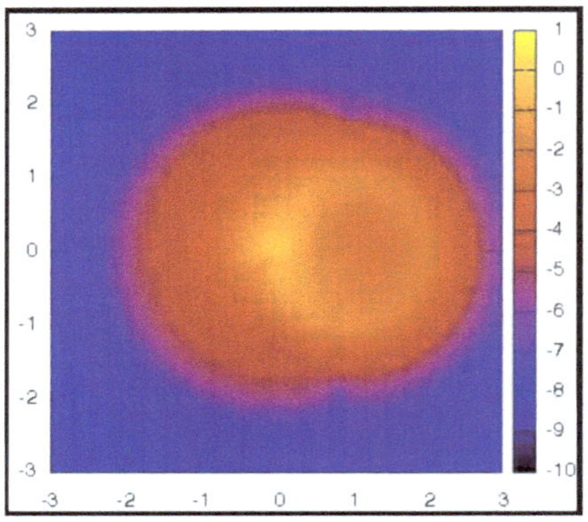

Figure 8: Contour plot of the electron momentum distribution computed in the impact-parameter first Born approximation (B1) for 400 keV proton impact on atomic hydrogen at b=1. The labeling is as for Fig. **7**.

$$\left(H_T - i\hbar\frac{\partial}{\partial t}\right)\psi^{B1}(\boldsymbol{r},t) = -V_p(\boldsymbol{r}-\boldsymbol{R}(t))\psi_i(\boldsymbol{r},t) \qquad \textbf{(6.16)}$$

It can also be evaluated indirectly using the inverse of the Fourier transform in Eq. (6.13),

$$a_{ki}^{B1}(\boldsymbol{b},K_{||}) = -i\left[\frac{(2\pi)^2}{v}\right](2\pi)^{-1}\int\exp(-i\boldsymbol{K}_{\perp}\bullet\boldsymbol{b})T_{ki}^{B1}(\boldsymbol{K})d^3K \qquad \textbf{(6.17)}$$

The factor $(2\pi)^2/v$ in Eq. (6.17) relates *T*-matrix elements to transition amplitudes [18] but has no effect on the momentum distribution.

Fig. **8** shows a contour plot of this amplitude computed for an impact parameter of 1 a.u. It is apparent that the momentum distribution at constant impact parameter is quite different from the momentum distribution at constant proton scattering angle. Of particular interest here is the apparent minimum very closed to the direct ionization maximum at the target location. This minimum is actually a zero of $a_{ki}(\boldsymbol{b})$ therefore by the imaging theorem it corresponds to a vortex in the time-dependent B1 wave function for infinite times. Since the vortex occurs in $P(\boldsymbol{b}, \mathbf{k})$, and the impact parameter is not the physically observed quantity, it is uncertain which distribution $P^{B1}(\boldsymbol{K}, \boldsymbol{k})$ or $P^{B1}(\boldsymbol{b}, \boldsymbol{k})$ should be observed. Actually, neither distribution corresponds to what is measured physically since both omit effects of $V_{PT}(R(t))$.

To include effects V_{PT} it is necessary to multiply $a_{fi}(\boldsymbol{b})$ by the phase factor $\exp[i\chi(\boldsymbol{b})]$ where

$$\chi(b) = \frac{1}{\hbar}\int_{-\infty}^{\infty} V_{PT}(R(t))dt = 2i\nu \log b \qquad \textbf{(6.18)}$$

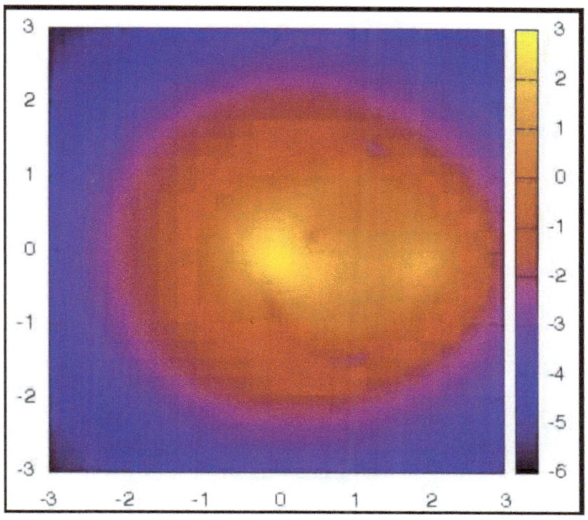

Figure 9: Contour plot of the electron momentum distribution computed in the eikonal Coulomb- Born approximation (B1) for 400 keV proton impact on atomic hydrogen at b=1. The labeling is as for Fig. **7**.

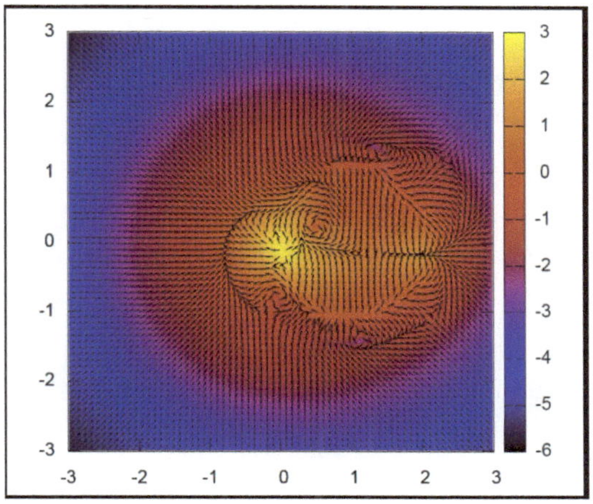

Figure 10: Same as Fig. **8** but with black arrows drawn in the direction of the velocity field.

and where $\nu = Z_P Z_T e^2/\hbar v$ is Somerfeld's parameter. This phase multiplies $a_{ki}(b)$ so it does not affect the impact parameter ionization probability $P(\boldsymbol{b}, \mathbf{k})$. As stated above, this probability is not the one that is measured, rather $P(\boldsymbol{K}, \boldsymbol{k}) = |a_{ki}(\boldsymbol{K})|^2$ is the observable quantity. The necessary amplitude $a_{ki}(\boldsymbol{K})$ is given by the Fourier transform of Eq. (6.17) with (6.18);

$$a_{ki}^{CB1}(\boldsymbol{K}_\perp, K_\parallel) = (2\pi)^{-1}\int b^{2iv} a_{ki}^{B1}(\boldsymbol{b})[-i\boldsymbol{K}_\perp \cdot \boldsymbol{b}]d^2b \qquad \textbf{(6.19)}$$

where the superscript CB1 designates that this is an approximation to the Coulomb-Born amplitude where incoming and outgoing relative Coulomb wave functions in the eikonal approximation are used.

The momentum distribution with CB1 is shown in Fig. **9**. Note that it differs considerably from the B1 distribution. In particular, there are holes in the CB1 distribution. To verify that these holes are vortices the plot is reproduced in Fig. **10** with small dark arrows drawn indicating the direction of flow of the velocity field at infinite times, $\boldsymbol{v} = \mathrm{Im}\nabla_k \log a_{ki}(\boldsymbol{K})$, for a value of $K_\perp$ given by the small angle scattering relation $K_\perp = 2Z_PZ_T/vb$. These holes are seen to relate to observable vortices.

At lower relative velocities v Somerfeld's parameter becomes large and the phase factor b^{2iv} oscillates rapidly. In this case a stationary phase evaluation of the CB1 amplitude is appropriate. Since the stationary phase point is at $b = 2Z_PZ_Te^2/v\hbar K_\perp$ one has that the b and K dependent electron momentum distributions are identical if the small angle scattering relation $b(\theta) = 2Z_PZ_T/v\theta$, where θ is the angle of scattering of the projectile, is employed to relate scattering angles to b. It is supposed that the small angle scattering relation holds for the momentum distributions in Fig. **3**. This enables the results of the imaging theorem as applied to the time-dependent Schrödinger equation Eq. (3.5) to be directly compared with experiment.

Notice that in the high energy regime where v < 1 and $b \approx 1$ as in Figs. **4-5** a simple relation between scattering angle an impact parameter does not accurately represent the eikonal CB1 electron momentum distribution and it is necessary to employ Eq. (6.19) without approximation. This suggests that angular momentum is transferred from the projectile to the target electron. If the final states are bound excited states, such transfer produces oriented states [35]. By oriented states we mean the mean value of the electron angular momentum operator

$\langle nl|\boldsymbol{L}|nl\rangle = \sum_{m,m'} \rho_{mm'} < nlm'|\boldsymbol{L}|nlm >$, where ρ is the density matrix of the nl state and m,m' are magnetic quantum numbers, does not vanish. For such bound states, the electron current rotates around the target nucleus. For continuum states, electron currents rotate in vortices about points where there is no center of force, and unexpected result which can none-the-less be tested experimentally.

Because the eikonal Coulomb-Born amplitude for proton impact ionization has vortex holes, this suggests that one look for such holes in electron momentum distributions computed in the exact Coulomb-Born approximation. Such distributions have been computed by Botero and Macek [36] for electron impact on the K-shell of carbon. They found dramatic differences between B1 and CB1 distributions of the ejected electrons but did not recognize the presence of vortices. Most notably, they found places where both the real and the imaginary part of the transition amplitude vanished as a function of the momentum vector of the ejected electron. The zeros almost certainly occur at vortices, but further study is needed to confirm this. If the zeros are indeed vortices their presence will certainly influence the interpretation of (e,2e) measurements on inner shells.

In this connection, Berakdar and Briggs [37, 38] examined a minimum observed by Murray and Reed [39] for electron impact on neutral He atoms. They used an approximate theory to compute the ionization amplitudes for the (e,2e) process and found an isolated point where both the real and the imaginary part of the amplitude vanished. In retrospect we can now identify this zero as a vortex, however, the vortex nature of the zero was confirmed by later calculations using the same approximate theory [32]. More accurate calculations [40] find an isolated zero that agrees well with experiment, possibly indicating that the experimental zero is actually a vortex.

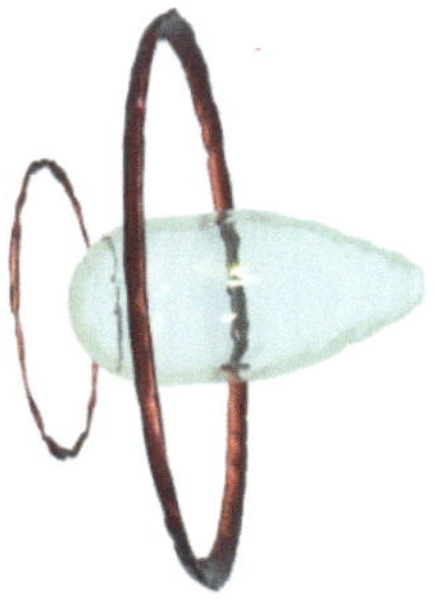

Figure 11: Three-dimensional plot of the time-dependent wave function for atomic hydrogen in a pulsed electric field showing two ring vortices.

The discussion of vortices has emphasized the role of angular momentum transfer in "stirring" the S-field to produce rotating electron current that leads to observable effects. The vortices thus formed fall into the class of line vortices that run from $-\infty$ to $+\infty$ along some curve. Net angular momentum (orientation [35]) is needed to form these line vortices. But Ref. [6] also discusses ring vortices that do not carry net angular momentum. These could, in principle, be present in the first Born amplitude B1 for ion impact ionization, however they are not found for the 1s states of hydrogen.

Ionization by linearly polarized electromagnetic fields could produce ring vortices. This possibility has been investigated [41] by using the LTDSE method and electric field pulses of 300 attoseconds (300 as) duration. The wave function after the pulse is turned off is illustrated in Fig. **11**. The two rings are regions where the magnitude of the wave function effectively vanishes, while the central lobe encloses a singly-connected region where the electron density is larger than a pre-assigned, but comparatively large, value. The two rings correspond to ring vortices showing that such structures are produced even without net transfer of vector angular momentum. This offers another avenue to study the effect of vortices in atomic processes.

VII. SUMMARY

In this chapter we have described how vortices are formed in time-dependent atomic processes. An imaging theorem is given which allows one to connect structures seen in electron momentum distributions $P(\boldsymbol{k})$ with structures in the time-dependent wave function for distances $\boldsymbol{r} = \boldsymbol{v}t$ and large times. Application of the RLTDSE method to one-electron atomic processes shows that vortices occur frequently in time dependent processes and can be studied experimentally. That vortices seen in actual calculations means that it now possible to bring the hydrodynamic picture of the Schrödinger field into close confrontation with experiment. Examples of collision and photon process that produce vortices are given.

ACKNOWLEDGEMENTS

Supported by the Office of Basic Energy Sciences, U.S. Department of Energy, through grants to the University of Tennessee (DE-FG02-02ER15283) and the Oak Ridge National Laboratory which is managed by UTBattelle, LLC under Contract No. DE-AC05- 00OR22725.

REFERENCES

[1] Born M. Zur quantenmechanik der stossvorgänge. Z Phys 1926; 37(11-12): 803-27.

[2] Macek JH, Sternberg J, Ovchinnikov SY, Lee TG, Schultz DR. Origin, evolution, and imaging of vortices in atomic processes. Phys Rev Lett 2009; 102(14): 143201.

[3] Lee TG, Ovchinnikov SYu, Sternberg J, *et al.* Quantum treatment of continuum electrons in the fields of moving charges. Phys Rev A 2007, 76(5): 050701(R).

[4] Dirac PAM. Quantised singularities in the electromagnetic field. Proc Roy Soc (Lond), 1931; A 133(821): 60-72.

[5] Takabayasi T. On the formulation of quantum mechanics associated with classical pictures. Progr Theor Phys 1952; 8(2): 143-82.

[6] Bialynicka-Birula I, Bialynicka-Birula Z, Sliwa C. Motion of vortex lines in quantum mechanics. Phys Rev A 2000; 61(3): 032110.

[7] Kleinpoppen H, Lohman B, Grum-Grzhimailo A, Becker U. Approaches to perfect/complete scattering experiments in atomic and molecular physics. Advances in Atomic, Molecular and Optical Physics 2005; 51: 471534.

[8] Dörner R, Khemliche H, Prior MH, *et al.* Imaging of saddle point electron emission in slow p-He collisions. Phys Rev Lett 1990; 77(22): 4520-3.

[9] Macek, JH, Ovchinnikov SYu. Theory of rapidly oscillating electron angular distributions in slow ion-atom collisions. Phys Rev Lett 1998; 80(11): 2298-01.

[10] Schultz DR., Strayer MR, Wells JC. Lattice, time-dependent Schrödinger equation solution for ion-atom collisions. Phys Rev Lett 1999; 82(20): 39769.

[11] Madelung O. Quantentheorie in hydrodynamischer form. Z Phys 1926; 40(3-4): 322-6.

[12] Schrödinger E. Quantisierung als eigenwertproblem. Ann Phys (Leipzig) 1926; 79: 361-76.

[13] Kiehn RM. An extension of Bohm's quantum theory to include non-gradient potentials and the production of nanometer vortices. University of Houston 1999, unpublished.

[14] Wigner EP. Relativistic invariance and quantum phenomena. Rev Mod Phys 1957; 29(3): 255-68.

[15] Smith FT. Did minkowski change his mind about noneuclidean symmetry in special relativity. Bulletin Am Phys Soc 2010; 55. Available from: http://meetings.aps.org/link/BAPS.2010.APR.W1.31.

[16] Mott NF. On the theory of excitation by collision with heavy particles. Proc Cambr Phil Soc 1931; 27(4): 553-60.

[17] Briggs JS, Rost JM. Time dependence in quantum mechanics. Eur Phys J D 2000; 10(3): 311-8.

[18] Briggs JS, Macek JH. The theory of fast ion-atom collisions. Adv At Mol Phys 1991; 28(1): 1-73.

[19] Fermi E. Quantum theory of radiation. Rev Mod Phys 1932; 4(1): 87-132.

[20] Gallaher DFL, Wilets L. Coupled-state calculations of proton-hydrogen scattering in the sturmian representation. Phys Rev 1968; 169(1): 138-49.

[21] Shakeshaft R. Coupled-state calculations of proton-hydrogen atom scattering using a scaled hydrogenic basis set. Phys Rev A 1978; 18(5): 1930-4.

[22] Winter T. Electron transfer, excitation, and ionization in p-H(1s) collisions studied with Sturmian bases. Phys Rev A 2009; 80(3): 032701.

[23] Bates DR, McCarroll R. Electron capture in slow collisions. Proc Roy Soc (Lond) 1958; A 245:175-83.

[24] Macek JH, Ovchinnikov SYu. Theory of rapidly oscillating electron angular distributions in slow ion-atom collisions. Phys Rev Lett 1998; 80(11): 2298-01.

[25] Solov'ev, EA. Transitions from a discrete level to the continuous spectrum upon adiabatic variation of the potential. Sov. Phys. JETP 1976; 43(3): 453-458.]; Nonadiabatic transitions in atomic collisions. Sov Phys Usp 1989; 32(3): 228-50.

[26] Wilets L, Wallace SJ. Eikonal method in atomic collisions. Phys Rev 1968; 169(1): 84-91

[27] Macek JH. Multichannel zero-range potentials in the hyperspherical theory of three-body dynamics. Few Body Systems 2002; 31(2-4): 241-8.

[28] Brauner M, Briggs JS, Klar H. Triply-differential cross sections for ionisation of hydrogen atoms by electrons and positrons. J Phys B: At Mol Opt Phys 1989; 22(14): 2265-87.

[29] Fetter AL, Walecka JD. Theoretical mechanics of particles and continua. New York: McGraw-Hill 1980.

[30] Hobsen A. Electrons as field quanta: A better way to teach quantum physics in introductory general physics courses. Am J Phys 2005; 73: 630-4.

[31] Briggs JS. Cusps, dips and peaks in differential cross-sections for fast three-body coulomb collisions. Comments At Mol Phys 1989; 23(1): 155-74.

[32] Macek JH, Sternberg JB, Ovchinnikov SY, Briggs JS. Theory of deep minima in (e,2e) measurements of triply differential cross sections. Phys Rev Lett 2010; 104(3): 033201.

[33] Landau LD, Lifshitz EM. Quantum mechanics, non-relativistic theory, 3rd ed. New York: Pergamon Press 1980; pp 628-9.

[34] Massey HSW, Mohr CBO. The collision of slow electrons with atoms. III. The excitation and ionization of helium by electrons of moderate velocity. Proc Roy Soc 1933; A 140(842): 613-36.

[35] Fano U, Macek JH. Impact excitation and polarization of the emitted light. Rev Mod Phys 1973; 45(4): 553-73.

[36] Botero J, Macek JH. Coulomb-Born calculation of the triple-differential cross section for inner-shell electron-impact ionization of carbon. Phys Rev A 1992; 45(1): 154-65.

[37] Berakdar J, Briggs JS. Three-body coulomb continuum problem. Phys Rev Lett 1994; 72(24): 3799-02.

[38] Berakdar J, Briggs JS. Interference effects in (e, 2e)-differential cross sections in doubly symmetric geometry. J Phys B: At Mol Opt Phys 1994; 27(18): 4271-80.

[39] Murray AJ, Read FH. Evolution from the coplanar to the perpendicular plane geometry of helium (e,2e) differential cross sections symmetric in scattering angle and energy. Phys Rev A 1993; 47(5): 3724-32.

[40] Colgan J, Al-Hagan O, Madison DH, *et al.* Deep interference minima in non-coplanar triple differential cross sections for the electron-impact ionization of small atoms and molecules. J Phys B: At Mol Opt Phys. 2009; 42(17): 171001.

[41] Schultz DR, Ovchinnikov SYu, Sternberg J. Private communication.

CHAPTER 2

Interatomic Electronic Decay Processes in Clusters

Vitali Averbukh[1,*], Lorenz S. Cederbaum[2], Philipp V. Demekhin[3], Simona Scheit[4], Přemysl Colorenč[5], Ying-Chin Chiang[2], Kirill Gokhberg[2], Sören Kopelke[2], Nikolai V. Kryzhevoi[2], Alexander I. Kuleff[2], Nicolas Sisourat[2] and Spas D. Stoychev[2]

[1]Department of Physics, Imperial College London, Prince Consort Road, London SW7 2AZ, United Kingdom; [2]Theoretische Chemie, Physikalisch-Chemisches Institut, Universitat Heidelberg, Im Neuenheimer Feld 229, D-69120 Heidelberg, Germany; [3]Institut fiir Physik, Experimental-Physik IV, Universitat Kassel, Heinrich-Plett-Str. 40, D-34132, Kassel, Germany; [4]Department of Basic Science, Graduate School of Arts and Sciences, The University of Tokyo, 153-8902, Tokyo, Japan and [5]Institute of Theoretical Physics, Faculty of Mathematics and Physics, Charles University in Prague, V Holešovičkách 2, 180 Prague, Czech Republic

Abstract: Since their theoretical prediction in 1997, Interatomic (intermolecular) Coulombic Decay (ICD) and related processes have been in the focus of intensive theoretical and experimental research. The spectacular progress in this direction has been stimulated both by the fundamental importance of the discovered electronic decay phenomena and by the exciting possibility of their practical application, for example in spectroscopy of interfaces. Interatomic decay phenomena take place in inner-shell-ionized and inner-shell-excited clusters due to electronic correlation between two or more cluster constituents. These processes lead to relaxation by electron emission and often also to disintegration of the resulting positively charged cluster. Here we review the recent progress in the study of interatomic decay phenomena in excited, singly and multiply charged clusters.

I. INTRODUCTION

The present day knowledge of interatomic (intermolecular) decay mechanisms in clusters encompasses a diversity of distinct physical phenomena, all stemming from interatomic (intermolecular) electronic interaction. In this section we give an overview of the predicted and observed interatomic decay processes induced by inner-shell ionization or excitation.

A. Interatomic (intermolecular) Coulombic Decay

Core vacancy states of atoms and molecules represent very highly excited states of the corresponding atomic or molecular ions, typically lying above the double or even multiple ionization thresholds. As a result, these states decay by electron emission in a specific type of autoionization process named after its discoverer, P. Auger [1]. Kinetic energies of the electrons emitted in the course of Auger decay are given by the differences between the bound states of singly and doubly charged species and thus are quantized. This property explains the great spectroscopic value of the Auger Electron Spectroscopy (AES) [2], as well as its importance for numerous analytical applications, *e.g.* in surface science (see, for example, Ref. [3]). Auger decay is typically an intraatomic process, only modestly affected by the environment. Usually, such an effect is manifested in the so-called chemical shift of the Auger electron lines (see, for example, Ref. [4]).

In 1997, the authors of the theoretical work [5] took a pioneering approach to the issue of the environment effects on the decay of vacancy states [5]. The question posed by the authors was:

Can a vacancy decay non-radiatively only due to the effect of the environment?

Surprisingly, it turned out that such an environment-mediated decay is not only possible, but is also a general phenomenon, typical of relatively low-energy inner-shell vacancies. In particular, clusters of

***Address correspondence to Vitali Averbukh:** Department of Physics, Imperial College London, Prince Consort Road London SW7 2AZ United Kingdom; Tel: +44 (0) 20 7594 7746; E-mail: v.averbukh@imperial.ac.uk

Gennadi Ogurtsov and Danielle Dowek (Eds)

various types prove to be the ideal objects to study this kind of decay phenomenon [5]. In order to get an idea of the new decay process discovered by Cederbaum and co-workers, one can consider the decay of 2s vacancy of neon, once in an isolated ion and once in a cluster, *e.g.* in Ne_n. The $2s^{-1}$ state of the isolated Ne^+ lies below the double ionization threshold of Ne and thus cannot decay by Auger mechanism. As a result, $(2s^{-1})$ Ne^+ decays radiatively on a nanosecond time scale. However, if $(2s^{-1})$ Ne^+ is allowed to interact with an environment, *e.g.* with other Ne atoms, the situation changes dramatically. Indeed, as shown schematically in Fig. **1** for neon dimer, in a Ne_n cluster, one can consider not only the high-energy Ne^{2+} Ne_{n-1} doubly ionized states, but also the ones of the type $(Ne^+)_2Ne_{n-2}$. The latter states are relatively low in energy due to the separation of the positive charge between two neon atoms. In fact, the charge-separated states lie several electronvolts lower than $(2s^{-1})$ Ne^+Ne_{n-1}. This leads to a very interesting *interatomic* decay process in which *2p* electron of the ionized Ne fills the 2s vacancy, while *2p* electron of another Ne atom is ejected into continuum. Since such a process is enabled by the Coulombic interaction between the electrons of the two Ne atoms, it has been called Interatomic Coulombic Decay (ICD). In a small loosely bound cluster, such as neon dimer, the repulsion between the two charges created by ICD leads to Coulomb explosion of the system [6] (see Ref. [7] for an exception). Under such conditions, the excess energy of the initial vacancy state is partitioned between the outgoing electron and the separating positively charged fragments. Thus, while Auger decay leads to quantized Auger electron energies, ICD in small clusters makes the total of the electron and the cluster fragment energies to be quantized. The kinetic energy of the relative motion of the fragments is often called Kinetic Energy Release (KER).

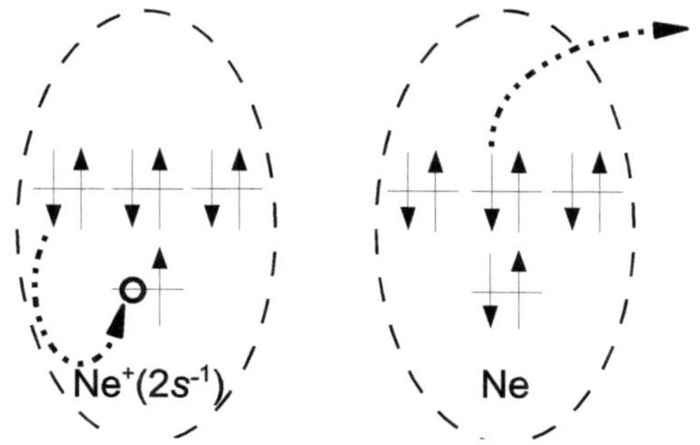

Figure 1: A scematic presentation of the ICD process in Ne dimer. 2p electron of the inner-valence-ionized Ne recombines into the 2s vacancy while a 2p electron of another Ne is ejected into continuum. The resulting doubly charged cluster decomposes by Coulomb explosion mechanism.

The last several years have witnessed a series of remarkable advances in the experimental study of ICD. Hergenhahn, Möller and co-workers have presented the first experimental evidence of ICD by clearly identifying the new process in neon clusters [8]. Dörner, Jahnke and co-workers have conducted a detailed study of ICD in neon dimer [9] using the Cold Target Recoil Ion Momentum Spectroscopy (COLTRIMS) [10]. They have been able to measure in coincidence both the ICD electrons and the neon ions generated by the Coulomb explosion of $(Ne^+)_2$. The coincidences detected by Frankfurt group were found to be arranged along the energy conservation line corresponding to the sum of the electron energy and the KER being about 5 eV. Thus, the experiment of Dörner and co-workers constitutes the most detailed direct proof of the ICD. The electron kinetic energy and the KER spectra of Frankfurt group were later confirmed by theoretical calculations [11]. Going back to larger neon clusters, Ohrwall *et al.* have estimated the dependence of the ICD lifetime on the neon cluster size by distinguishing between the "bulk" and the "surface" peaks in the ICD electron spectra [12]. These experimental findings were found to be in a reasonable agreement with earlier theoretical predictions of Santra *et al.* [13] (see also the more recent theoretical work of Vaval and Cederbaum [14]).

Both theoretical and experimental investigations have established ICD as a highly general and a very efficient decay process. Indeed, ICD is characteristic of vacancy states of Van der Waals clusters (see, *e.g.* Refs. [5, 8]), hydrogen bonded clusters (see, *e.g.* Refs. [15, 16]), and even endohedral fullerenes [17]. The ICD lifetimes were found to belong to the range of 1 to 100 fs [12, 13, 17], many orders of magnitude shorter than those of the competing photon emission process. Thus, ICD is the main decay mode of moderate-energy (Auger-inactive) inner shell vacancies in clusters. Further studies of ICD are motivated, however, not only by the generality and efficiency of this new physical process, but also by the perspectives of its practical use, for example, in spectroscopy. The very first step in this highly promising direction has been already done by Hergenhahn and co-workers who have shown that ICD electron spectra can be used in order to identify the Ne/Ar interface [18].

B. Beyond ICD of Singly Ionized States

1. Interatomic Decay of Inner-Shell Excitations

Recently, Barth *et al.* [19] have addressed the question, whether interatomic decay can occur not only in the inner-valence-ionized, but also in the inner-valence-excited states of clusters. They have created Ne $(2s^{-1}3p)$ excitations in Ne_n clusters (n being 70 on average) and detected the electrons emitted due to the $(2s^{-1}3p)NeNe_{n-1} \rightarrow (2p^{-1}3p)Ne(2p^{-1})Ne^+Ne_{n-2} + e^-$ process. Aoto *et al.* [20] studied in detail a similar decay phenomenon in neon dimer. This process is related to ICD exactly in the same way in which the resonant Auger effect [21, 22] is related to the regular Auger effect [1, 2]. Consequently, it has been called Resonant Interatomic Coulombic decay (RICD) [19].

RICD physics is richer and more involved than the ICD physics due to several reasons. First, the interatomic decay of inner-shell-excited states is accompanied by the intraatomic autoionization, *e.g.* $(2s^{-1}3p)Ne \rightarrow (2p^{-1})Ne + e^-$. Whereas ICD competes only with slow radiative decay, RICD has to compete with a fast non-radiative process. Nevertheless, both experimental [19, 20] and theoretical [23] evidence show that this competition does not lead to a suppression of RICD. Another important difference between ICD and RICD comes from the fact that the inner-valence-excited electron can participate in RICD process. Exactly as the resonant Auger decay [21, 22], RICD can occur either by spectator (sRICD) or by participator (pRICD) mechanism. While the sRICD process has been observed experimentally, pRICD has yet to be identified in the RICD electron spectra.

Another distinction between ICD and RICD has its origin in the higher energy accumulated in the inner-valence-excited states relative to the one of the inner-valence-ionized states. For example, $(2s^{-1}3p)Ne$ lies about 45.5 eV above the Ne ground state, whereas $(2s^{-1})\ Ne^+$ lies only about 26.9 eV above the Ne+ ground state. As a result, decay of inner-valence-excited states can be accompanied by double ionization of the cluster. This can happen according to a variety of mechanisms, which have been discussed qualitatively in Ref. [23]. The predicted double ionization interatomic processes still await their detailed quantitative study. The essential question is whether the double ionization processes are fast enough to compete with autoionization and sRICD.

2. Auger-ICD Cascade

It is well known that Auger decay of core vacancies often results in highly excited states of the corresponding doubly ionized species. Sometimes, this brings about another stage (or even several stages) of Auger decay, forming what is usually called a decay cascade. Often, however, the excited doubly ionized states created by Auger process are not energetic enough to decay by an intraatomic mechanism.

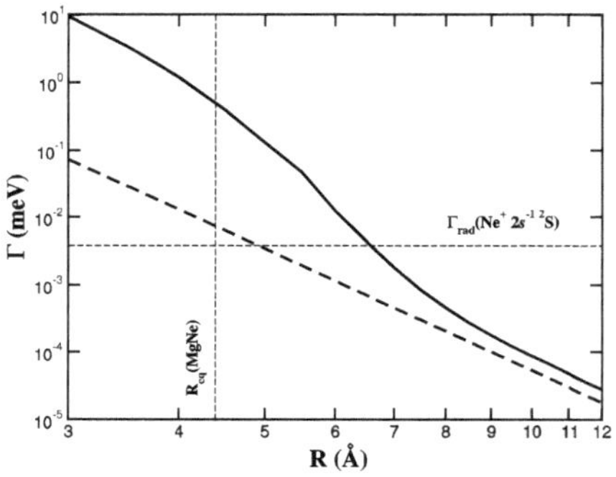

Figure 2: Schematic representation of ETMD process.

Under such conditions, formation of decay cascade is impossible in isolated species, but can occur in clusters with the second step of the cascade being of the ICD, rather than of the Auger type. The Auger-

ICD cascade has been first predicted by Santra and Cederbaum [24] in 1s-ionized neon dimer and has first been observed by Ueda and co-workers [25-27] in 2p-ionized argon dimers. Interestingly, the ICD process following the $2p^{-1} - 3s^{-1}3p^{-1}$ Auger transition in argon dimer is energetically forbidden: the $3s^{-1}3p^{-1}$ states of Ar^{2+} are not energetic enough to lead to ICD [28]. Observation of Auger-ICD cascade in Ar_2 [25-27] has been nevertheless possible due to the fact that Auger decay populates not only the $3s^{-1}3p^{-1}$ main states, but also higher-energy satellites having admixture of $3p^{-3}3d$ configurations. The ICD following 2p ionization of the Ar dimers was theoretically interpreted in [28]. Later on, Auger-ICD cascade was also observed following 1s ionization of Ne dimer [29-31]. The ICD process was assigned as transition between the $(2s^{-1}2p^{-1}\ {}^1P)Ne^{2+}Ne$ and $(2p^{-2}\ {}^1D)Ne^{2+}(2p^{-1})Ne^+$, and interpreted theoretically in [32]. More recently, relaxation pathways of the $Ne^{2+}Ar$ states populated *via* the K-LL Auger decay of $(1s^{-1})\ Ne^{2+}Ar$ were analyzed in detail in [33]. Experimental work on the Auger-ICD cascade in NeAr is now in progress [34].

3. Electron transfer Mediated Decay

Fig. **1** implies that ICD is mediated by energy transfer between the ionized and the neutral cluster units. One can also imagine, however, an interatomic decay process mediated by electron transfer. Indeed, such a mechanism is presented schematically in Fig. **2**. It is analogous to the well-known Transfer Ionization (TI) in collisions [35] and has been first predicted for clusters in Ref. [36]. The energetic condition that needs to be fulfilled for the ETMD to become operational is that the inner or outer shell ionization energy of one atom or molecule in a cluster (Ne in Fig. **1**) should exceed the double ionization energy of a neighboring atom or molecule (Ar in Fig. **1**). In the case of the inner shell ionization, ETMD is usually suppressed by the much faster ICD [36]. In the case of the outer shell ionization, however, ETMD can turn out to be the main decay channel. The latter scenario is realized, for example, in miclrosolvation clusters of Li^+ [37]:

$$Li^{2+}H_2O \rightarrow Li^+\ H_2O^{2+} \qquad \textbf{(1)}$$

Interestingly, energy transfer and electron transfer mechanisms can be combined in a three-center decay process, whereby the energy released in the hole filling by a water electron is used to ionize a neighboring water molecule [37]:

$$Li^{2+}\ (H_2O)_2 \rightarrow Li^+(H_2O^+)_2 \qquad \textbf{(2)}$$

Very recently, the basic two-center ETMD process has been measured experimentally in mixed clusters of Ar and heavier noble gases [38].

Further exploration of the fascinating subject of the interatomic decay phenomena and development of spectroscopic tools on their basis requires intensive theoretical effort to guide the experimental work. Such an effort is hardly possible without efficient, advanced theoretical tools involving both *ab initio* description of the electron correlation driving the decay and a treatment of the ensuing dynamics of the ionized cluster fragments. The next section gives the theoretical picture of interatomic decay within the Born-Oppenheimer (BO) approximation. *Ab initio* theory of interatomic decay widths is presented in some detail for the case of the ICD process in Section III. Section IV is devoted to the theory of interatomic decay of doubly ionized states applied to Auger-ICD cascades and to the collective decay of two inner-shell vacancies. The state of the art of the theory of RICD is given in Section V. Some considerations on the future of the field are summarized in Section VI.

II. COUPLED ELECTRONIC AND NUCLEAR DYNAMICS OF INTERATOMIC DECAY

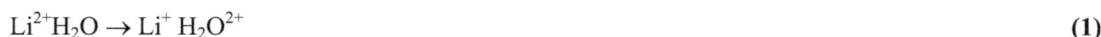

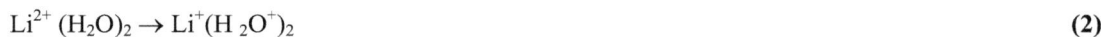

The main objective of the theory of ICD is efficient and reliable calculation of the measurable spectra, *i.e.* ICD electron kinetic energy spectrum and (where applicable) KER spectrum. The theoretical description is usually given within Born-Oppenheimer approximation, in which the electronic states are decoupled from nuclear motion and depend only parametrically on the nuclear coordinates. In this picture, the inner shell ionization and the subsequent ICD process can be visualized as a series of transitions between Potential Energy Surfaces (PESs) belonging to electronic states of different number of electrons (*i.e.* accompanied by electron emission). These transitions are represented schematically in Fig. **3**. Initially, the system is assumed to be in the ground electronic state of the neutral (N-electron) system. The corresponding PESs of loosely bound clusters are characterized by shallow minima (*e.g.*, in meV range for Van der Waals systems) and large equilibrium interatomic distances. Photoionization brings the cluster almost instantaneously into inner-shell-ionized (typically, inner-valence-ionized) [(N — 1)-electron] state, being the intermediate state of the decay The PES of the singly ionized system is affected by the charge - induced dipole interaction that increases the binding energy and decreases the equilibrium interatomic distances relative to the Van der Waals ground state. This means that after landing on the inner-shell-ionized PES, the nuclear wave packet is driven towards shorter internuclear distances. Due to the ICD, the intermediate state has finite lifetime. This means that the nuclear wave packet moving on the intermediate state PES can lose some of its density.

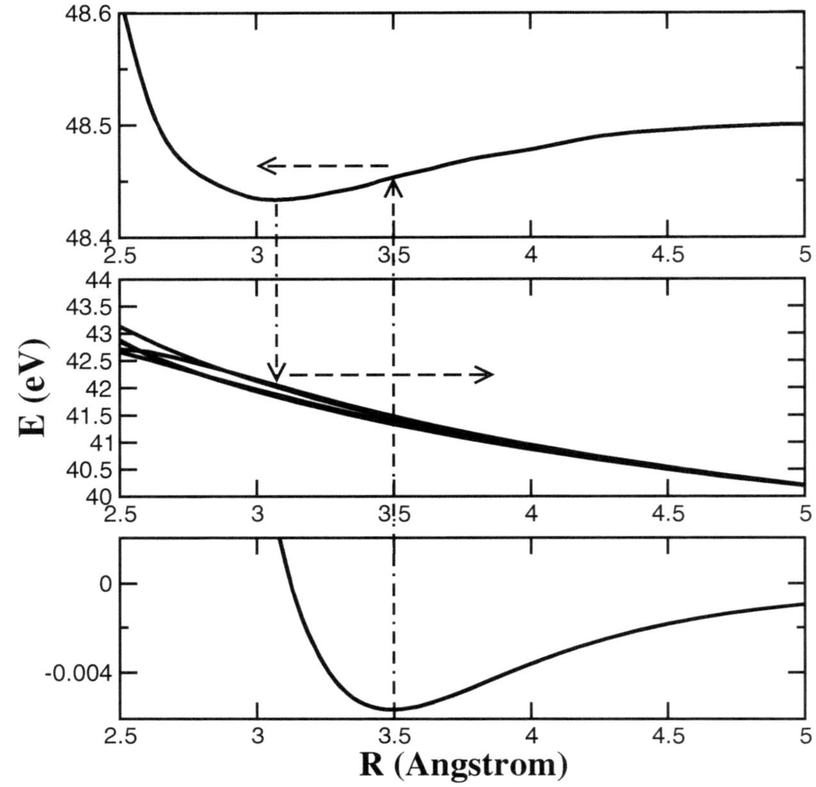

Figure 3: Schematic representation of ICD process in Born-Oppenheimer picture. PESs of NeAr [40] are used as a representative example. Lower frame: ground state PES of the neutral diatom (initial state). Upper frame: inner-valence-ionized (intermediate state) PES. Middle frame: doubly ionized (final state) pESs. Transitions between PESs accompanied by loss of an electron are shown by dashed-dotted lines. Directions of motion of nuclear wave packets on intermediate and final state PESs are shown by dashed lines.

to the final (doubly ionized) state PESs. The latter are typically dominated by the repulsion between the two positive charges and are often close to the purely Coulombic repulsive shape. The geometry at which the decay occurred determines the partition of energy between the outgoing electron and the repelling fragments.

The above qualitative picture has its full formal analog in the so-called time-dependent formulation of the theory of nuclear dynamics of electronic decay processes. This theory is given in some detail in Ref. [11] and in references therein. Here we will consider only its few principal points. Let us denote the nuclear wave packets for the initial (*i*), intermediate (*d*) and final (*f*) electronic states as Ψ_i, Ψ_d, and Ψ_m^f, respectively, where the index *m* accounts for the possibility that there are several final electronic states (see Fig. **3**). We assume that the electric field used in order to ionize the initial state is weak enough (weak field approximation) and describe the coupling between the intermediate and final states of the decay in the so-called local approximation [11]. Under these approximations, the nuclear wave packets obey the following set of differential equations:

$$i\hbar\frac{\partial}{\partial t}\left|\Psi_i(R,t)\right\rangle = \hat{H}_i(R)\left|\Psi_i(R,t)\right\rangle$$
$$i\hbar\frac{\partial}{\partial t}\left|\Psi_d(R,t)\right\rangle = \hat{F}(R,t)\left|\Psi_i(R,t)\right\rangle + \hat{H}_d(R)\left|\Psi_d(R,t)\right\rangle \quad \textbf{(3)}$$
$$i\hbar\frac{\partial}{\partial t}\left|\Psi_m^f(R,\varepsilon,t)\right\rangle = \hat{W}_m(R,\varepsilon)\left|\Psi_d(r,t)\right\rangle + (\hat{H}_m^f(R)+\varepsilon)\left|\Psi_m^f(R,\varepsilon,t)\right\rangle$$

where ε is the energy of the electron emitted during ICD.

The Hamilton operators for the nuclear motion in the initial and final electronic states in (3) are defined as $\hat{H}_i = \hat{T}_N + \hat{V}_i$ and $\hat{H}_m^f = \hat{T}_N + \hat{V}_m^f$, where T_N is the nuclear kinetic energy and V_i and V_m are the initial and final state PESs, respectively. The effective Hamilton operator governing the intermediate state dynamics has to account for the fact that the intermediate state is an electronic resonance. Within the local approximation, this is done by lending the intermediate state Hamiltonian a non-Hermitian character:

$$\hat{H}_d(R) = \hat{T}_N(R) + \hat{V}_d(R) - i\Gamma(R)/2 \quad \textbf{(4)}$$

where $V_d(R)$ is the intermediate state PES and $\Gamma(R)$ is the total decay width.

The coupling operators F and W_m describe the excitation from the initial to the intermediate electronic state and the coupling of the latter to the mth final state, respectively. Since we assume the inner-shell ionization to occur instantaneously, F can be taken to be R-independent and simply a δ-function in time. The W_m operators describe the coupling of the intermediate state to the different final states. Within the local approximation, they are energy-independent and are related to the corresponding partial decay widths: $\Gamma(R) = 2\pi\left|\hat{W}_m(R)\right|^2$.

As shown, *e.g.*, in Ref. [11, 38, 39], all the information concerning the decay spectrum can be derived from the knowledge of the final nuclear wave packets. Indeed, at sufficiently large time (*i.e.*, when the decay is complete and the norm of all the intermediate wave packet is zero), only the final states are populated and thus carry all the spectroscopic information of interest. In particular, the decay spectrum as a function of the emitted electron energy ε is given by [11, 38].

$$\sigma(\varepsilon) = \lim_{t\to\infty}\sum_m \sigma_m(\varepsilon,t) = \lim_{t\to\infty}\sum_m \left\langle \Psi_m^f(R,\varepsilon,t) \middle| \Psi_m^f(R,\varepsilon,t)\right\rangle \quad \textbf{(5)}$$

Examples of the ICD electron spectra calculated for a series of initial vibrational states of the neutral NeAr cluster [40] are given in Fig. **4**.

Eqs. (3) show clearly that the shape of the spectrum (5) is determined by two competing time scales: that of the nuclear wave packet motion on the intermediate PES and that of the interatomic decay. If the decay is fast relative to the nuclear motion, the ICD occurs around the equilibrium geometry of the neutral (see Fig. **4**) and the ICD electron spectrum has a simple shape related to the structure of the initial vibrational state of the neutral cluster. If, on the other hand, ICD occurs within a wide range of geometries because of the long ICD lifetime or because of the vast spread of the vibrational ground state of the neutral, various contributions to the ICD electron spectrum are expected to interfere resulting in a complex pattern of σ(E) [see Eq. (5)]. The latter situation was predicted for unrealistic parameters in Ref. [6] and has been recently realized experimentally in helium dimer [41], where ICD is possible due to a simultaneous ionization and excitation of one of the He atoms (see also Ref. [42]). An alternative way to obtain the ICD electron spectrum is through the time-independent formulation. The time-independent theory uses the stationary vibrational eigenstates of the intermediate and final state PESs $\left|n_k^d\right\rangle$ and $\left|n_j^{f_m}\right\rangle$, to express the ICD spectrum:

$$\sigma(\varepsilon) = \sum_j \left| \sum_k \frac{(n_j^{f_m} | \hat{W}_m^* | n_k^d)(n_k^d | F | n_0)}{\varepsilon + E_j^{f_m} - E_n^d} \right|^2 \tag{6}$$

(exclusive population of the vibrational ground state PES is assumed here). Eq. (6) can be derived using non-Hermitian quantum scattering theory (see, *e.g.*, Ref. [43]). It closely resembles the well known Kramers-Heisenberg formula [44] used, for instance, for description of resonant X-ray emission (see Ref. [45] for an example of a physical situation where Kramers-Heisenberg-type expression is not applicable to the Auger/ICD electron spectrum). The important difference between Eq. (6) and the standard Kramers-Heisenberg formula stems from the fact that the vibrational states in Eq. (6) are eigenstates of the *non-Hermitian* Hamilton operator H_d (4) describing the intermediate resonant electronic state. As a result, one has to use the so-called c-product (-|-) [see Eq. (6)] which is defined by (f |g) = <f*|g> (see Ref. [46]). Clearly the time-independent and time-dependent formulations are absolutely equivalent [47], even though each of them gives a different insight in the spectrum. For instance, in the time-independent expression the role of the single vibrational states can be visualized. On the other hand, the time-dependent formalism allows to follow the time-evolution of the process and of the electron distribution, often simplifying the interpretation of the final spectrum. Clearly, an accurate computation of the ICD width, Γ(*R*), of the intermediate electronic state is crucial for a reliable prediction of the ICD electron spectra within both time-dependent and time-independent approaches.

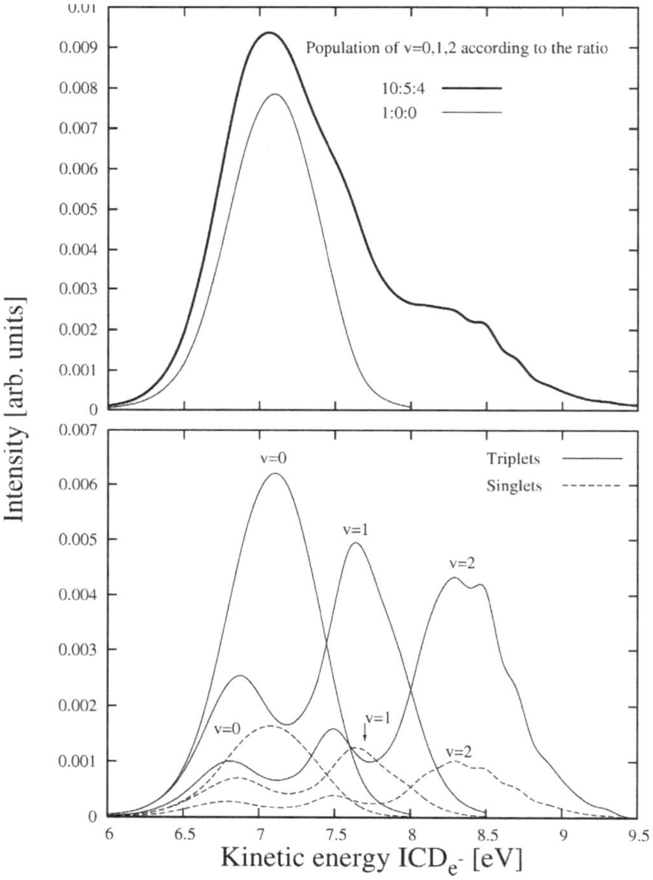

Figure 4: ICD of 2**s** Ne vacancy in NeAr cluster [40]. Lower panel: the singlet (dashed lines) and the triplet (solid lines) final state contributions to the ICD electron spectra for three different initial vibrational states of neutral NeAr: **v** = 0, **v** = 1, and **v** = 2. Upper panel: the total ICD electron spectra obtained assuming the population of the three lowest vibrational states of the electronic ground state to be according to the 10 : 5 : 4 ratio (solid line) and according to the 1 : 0 : 0 ratio (thin solid line). One can see that the ICD electron spectrum depends strongly on the initial vibrational state of the neutral cluster.

III. ICD WIDTHS BY FANO-ADC METHOD

Calculation of ICD widths can be achieved within one of the two main theoretical approaches. One of them relies on the introduction of complex absorbing potential (CAP) [48, 49] into the (N — 1)-electron Hamiltonian [50] (see Ref. [51] on the relation between the CAP method and the method of exterior complex scaling). The decay widths are then given by the imaginary parts of those eigenvalues of the resulting non-Hermitian Hamiltonian that are stationary with respect to the non-physical CAP parameters. The (N — 1)-electron Hamiltonian can be represented using a variety of *ab initio* techniques, such as for example, configuration Interaction (CI) or Algebraic Diagrammatic Construction (ADC) [52]. The application of the CAP-CI method to calculation of ICD widths has been reviewed in detail in Ref. [53]. More recently, CAP-ADC method [53] based on the ADC representation of the many-electron Hamiltonian has been developed [54, 55] and applied to ICD [14]. An alternative *ab initio* approach to calculation of the interatomic decay widths on which we would like to concentrate here [56] relies on Fano theory of resonances [57], ADC-ISR representation of the many-electron wave functions [58] and Stieltjes imaging technique [59].

A. ICD within Fano Theory of Resonances

Fano theory of resonances [57] as well as its generalized version [60, 61] developed for the description of Auger decay widths represents the wave function $\mathbf{\Psi}_E$ at some energy E above threshold as a superposition of bound-like (Φ) and continuum-like (χ_ε) components, which can be thought of as the initial and final states of the decay.

$$\Psi_{\alpha,E} = a_\alpha(E)\Phi + \sum_{\beta=1}^{N_c}\int c_{\beta,\alpha}(E,\varepsilon)\chi_{\beta,E}d\varepsilon \qquad (7)$$

(7) where the index β runs over the N_c possible decay channels. In the specific case of interatomic (intermolecular) decay in clusters, the bound part of the wave function, Φ, corresponds to the singly ionized state, typically created by the inner valence ionization of one of the cluster subunits. The state Φ is characterized by the mean energy:

$$E_\Phi = \langle\Phi|H|\Phi\rangle \qquad (8)$$

H being the full Hamiltonian of the system. The N_c decay channels in Eq. (7) are defined by the doubly ionized states of the cluster characterized by the energies, $E_\beta < E$, $\beta = 1,., N_c$, *i.e.* by the energetically accessible final states of the interatomic (intermolecular) decay. The continuum functions corresponding to the decay channels are assumed to diagonalize the Hamiltonian to a good approximation:

$$\langle\chi_{\beta',\varepsilon'}|H-E|\chi_{\beta,\varepsilon}\rangle \approx (E_\beta+\varepsilon-E)\delta_{\beta,\beta'}\delta(E_{\beta'}+\varepsilon'-E_\beta-\varepsilon) \qquad (9)$$

Using the assumption of uncoupled continuum functions, Fano theory provides an analytic expression for the evaluation of the decay width:

$$\Gamma = \sum_{\beta=1}^{N_c}\Gamma_\beta = 2\pi|M_\beta(E_r,\varepsilon_\beta)|^2, \quad M_\beta(E,\varepsilon) = \langle\Phi|H-E|\chi_{\beta,\varepsilon}\rangle \qquad (10)$$

where E_r is the real energy of the decaying state, $E_r \approx E_\Phi = \langle\Phi|H-E|\Phi\rangle$ and ε_β is the asymptotic kinetic energy of the ejected electron for the decay channel β, $E_r = E_\beta + \varepsilon_\beta$.

B. Initial and Final States of the ICD by Algebraic Diagrammatic Construction in the Framework of the Intermediate State Representation

For the result (10) to be applicable to the computation of the interatomic decay rates, one has to provide sensible approximations for the multi-electron bound (Φ) and continuum ($\chi_{\beta,\varepsilon}$) wave functions. In our case, these are wave functions of a singly ionized N-electron cluster, *i.e.* (N - 1)-electron states. Such states can be conveniently constructed using the single ionization ADC technique. The ADC methodology has been originally developed within the Green's function formalism [52]. Here, however, we would like to briefly review the single ionization ADC from a different standpoint, using the Intermediate State Representation (ISR) as proposed by Schirmer [58].

Consider the Hartree-Fock (HF) ground state of the N-electron neutral cluster, Φ_0^N. One can form a complete orthonormal set of the (N - 1)-electron basis functions, $\Phi_J^{(N-1)}$ applying the so-called physical excitation operators, $\{C\mathbf{j}\}$, to the HF ground state:

$$\Phi_J^{(N-1)} = \hat{C}_J \Phi_0^N \quad \left\{\hat{C}_J\right\} \equiv \left\{c_i; c_a^* c_i c_j, i < j; c_a^* c_b^* c_i c_j c_k, a < b, i < j < k; ...\right\} \tag{11}$$

where c_i and c_a^* are annihilation and creation operators respectively, the subscripts *i, j, k*,. relate to the occupied spin-orbitals and the subscripts *a, b, c*,... relate to the unoccupied spin-orbitals. The basis set (11) is used in the familiar CI expansion of the wave function. This expansion, once truncated after some specific excitation class [J], possesses such important drawbacks as slow convergence and lack of size-consistency. The ADC method overcomes these drawbacks by using a more complicated basis for the expansion of the (N - 1)-electron wave functions. The idea is to apply the physical excitation operators, $\{C_J\}$, to the perturbation-theoretically corrected, or "correlated" ground state of the neutral system,

$$\Psi_J^0 = C_J \Psi_0^N \quad \Psi_0^N = \Phi_0^N + \Psi_0^{(1)} + \Psi_0^{(2)} + \Psi_0^{(3)} + ... \tag{12}$$

where $\Psi_0^{(n)}$ is the nth-order correction to the HF ground state obtained by the standard many-body perturbation theory (see, *e.g.* Ref. [62]). Unfortunately, the resulting Correlated Excited States (CES's), Ψ_J^0, are not orthonormal. ADC takes care of this problem by orthonormalizing them in two steps to obtain the so-called intermediate states, Ψ_J. First,Gram-Schmidt orthogonalization *between the excitation classes* is performed to obtain the "precursor" states, $\Psi_J^{\#}$:

$$\Psi_J^{\#} = \Psi_J^0 - \sum_{\substack{K \\ |K|<|J|}} \left\langle \tilde{\Psi}_K \middle| \tilde{\Psi}_J^0 \right\rangle \tilde{\Psi}_K \tag{13}$$

i.e. the functions belonging to the higher [*e.g.* two-hole, one-particle (2h1p) or [J] = 2] excitation class are made orthogonal to those of all the lower [in this case, only one- hole (1h) or [K] = 1] excitation classes. Second, the precursor states are orthonormalized *symmetrically* inside each excitation class:

$$\tilde{\Psi}_J = \sum_{\substack{J' \\ |J'|<|J|}} (\rho^{\#-\frac{1}{2}})_{J',J} \tilde{\Psi}_{J'} \quad (\rho^{\#})_{J,J'} = \left\langle \Psi_{J'}^{\#} \middle| \Psi_J^{\#} \right\rangle \tag{14}$$

where $\underset{=}{(\rho^{\#})_{J,J'}}$ is the overlap matrix of the precursor states belonging to the same excitation class. The above two-step procedure can be applied iteratively, noting that the correlated excited states of the lowest (1h) excitation class are by definition also the precursor states.

Any state of the (N — 1)-electron system can be represented using the orthonormal basis of the intermediate states.

$$\Psi_{\ddot{u}}^{(T-1)} = \sum_i \sum_{|J|=i} Y_{q,J} \tilde{\Psi}_J \tag{15}$$

The expansion coefficients, Y_J are obtained by the diagonalization of the Hamiltonian matrix constructed in the basis of the intermediate states. It is a crucial feature of the ADC approach that the Hamiltonian matrix elements of the type $\langle \tilde{\Psi}_J | \tilde{\Psi}_J \rangle$ can be expressed analytically *via* the orbital energies and the electron repulsion integrals if one performs the orthonormalization procedure of Eqs. (13,14) approximately and consistently with the order of the many-body perturbation theory which is used for the construction of the correlated ground state [see Eq. (12)]. Moreover, it can be shown [58] that truncation of the expansion (15) after the excitation class [J] = m introduces an error of the order of 2m, which should be compared to m +1 for the slower-converging CI expansion. The accuracy of the expansion in excitation classes (15) should be, of course, consistent with that of the perturbation theoretical series for the correlated ground state (12). Thus, the order, n, at which the perturbation theoretical expansion (12) is truncated is the single parameter defining the level of the ADC approximation. For this reason, ADC schemes of various quality are usually denoted as ADC(n), n=2, 3, 4,. in full analogy with the well-known MP2, MP3, MP4. perturbation-theoretical techniques for the ground state of the neutral system.

The ADC(2) scheme for singly ionized states describes the many-electron wave-functions in the basis of 1h and 2h1p intermediate states treating the coupling between the 1h states and between 1h states and 2h1p states in the second and in the first order, respectively. ADC(2) approximation neglects the coupling between the different 2h1p basis functions. The extended ADC(2) scheme [ADC(2)x] takes into account the coupling between the 2h1p states in the first order (*i.e.* on CI level). The third-order ADC(3) scheme, while still confined to the basis of 1h and 2h1p intermediate states, treats the coupling between the 1h states and between 1h states and 2h1p states in the third and in the second order, respectively. A detailed description of the single ionization ADC(2) and ADC(3) schemes, including the expressions for the Hamiltonian matrix elements can be found in Ref. [63]. The proof of the size-consistency of the ADC(n) schemes has been given in Ref. [58]. The main limitation of the existing ADC(n) schemes is that they are applicable to ionized and/or excited states of closed-shell systems only. Here we are interested in applying ADC to the interatomic (intermolecular) decay in ionized Van der Waals and hydrogen bonded clusters, all of which satisfy this requirement.

Our main purpose is to demonstrate that the ADC(n) schemes can be used for the *ab initio* calculations of the decay rates within Fano formalism. To this end, we need to show that both bound (Φ) and continuum $\chi_{\beta,\varepsilon}$ components of the (N — 1)-electron wave function describing the decay process [see Eq. (7)] can be approximated by the expansion in the basis of the intermediate states (15). Suppose, a vacancy residing on the subunit A of a weakly bound cluster can decay by one of the interatomic (intermolecular) mechanisms, but cannot decay non-radiatively if created in the isolated species A. Clearly, in the final state of such a decay will be characterized by two vacancies, one or both of them residing on another cluster subunit. Thus, in order to construct the ADC(n) approximation for the bound part, Φ, one can restrict the physical excitation operators of the higher excitation classes to such where all the holes reside on the subunit A only:

$$\Psi_J^0 = C_J \Psi_0^N \quad \{\hat{C}_J\} \equiv \{c_i; c_a^* c_i c_j, i<j, \phi_{i,j} \in A; c_a^* c_b^* c_i c_j c_k, a<b, i<j<k, \phi_{i,j,k} \in A; \ldots\} \tag{16}$$

where $\phi_{ııı} \in A$ is an occupied spin-orbital of the neutral cluster localized on the subunit A. In this way, the intraatomic (intramolecular) relaxation and correlation effects inside the subunit A are taken into account, whereas any kind of interatomic decay cannot be described due to the restriction imposed on the holes. Upon the completion of the selection process, one can construct and diagonalize the Hamiltonian in the basis of the restricted set of the intermediate states using the standard methods. The ADC(n) state approximating the Φ component can be identified, for example, as the one possessing the maximal overlap with the cluster orbital representing the initial vacancy. Since no configurations corresponding to the open decay channels were used in the ADC-ISR expansion for the bound-like component, Φ, will be one of the lowest-energy eigenvectors of the ADC Hamiltonian. Therefore, a highly efficient Davidson diagonalization technique [64] can be used to diagonalize the matrix.

Once the ADC(n) approximation for the bound component of the wave function has been provided, the remaining task is to construct the approximate continuum components, $\chi_{\beta,\varepsilon}$ describing the possible final states of the interatomic (intermolecular) decay. Such states are naturally found among the ADC(n) eigenstates of the 2h1p character:

$$\chi_{\beta,\varepsilon} \propto \Psi_q^{2h1p} = \sum_i \sum_{|J=i|} Y_{q,i} \hat{\Psi}_J, \quad 1 - \sum_{|J|=2} \left| Y_{q,J} \right|^2 << 1 \tag{17}$$

The Ψ_q^{2h1p} functions can be constructed without any restriction of the kind of (16) being imposed on the physical excitation operators. It is, thus, possible that some intermediate states contribute both to Φ and to Φ_q^{2h1p} expansions, leading to $\langle \Phi | \Psi_q^{2h1p} \rangle \neq 0$. This does not lead to complications as the Fano formalism that we are using does not assume the orthogonality of the bound and the continuum components. Application of the selection scheme described above is not straightforward in the case of symmetric clusters, *e.g.* those in which the ionized inner-shell orbital is delocalized due to inversion symmetry. As has been shown in Ref. [65], this difficulty can be circumvented by using the appropriate linear combinations of the "standard" configurations.

Despite the ability of ADC(n) to produce 2h1p-like wave functions in the continuum region of the spectrum, there still exists a major difficulty in associating these ADC(n) eigenstates with the approximate continuum states of Fano theory. The difficulty stems from the fact that the ADC(n) calculations, and *ab initio* quantum chemical calculations in general, are routinely performed using the L^2 bases, usually the Gaussian ones. As a result, the L^2 and not the scattering boundary conditions are imposed and the Φ^{2h1p} functions are not properly normalized,

$$\langle \Psi_{\breve{u}}^{2pl_3} | \Psi_{\breve{u}'}^{2pl_3} \rangle = \delta_{q,q'} \tag{18}$$

[compare to Eq. (9)]. Moreover, the corresponding eigenenergies, E_q^{2h1p} are discrete and are not expected to fulfil the energy conservation relation for the non-radiative decay, $E_q^{2h1p} = E_\Phi$, except by a coincidence. An efficient way to deal with the above complications is provided by the computational approach known as Stieltjes-Chebyshev moment theory or Stieltjes imaging [59]. This approach rests on the fact that while decay width (10) can not be calculated using the discretized continuum functions directly, the spectral moments of (10) calculated using the pseudospectrum are good approximations to the spectral moments constructed using the true continuum [59]:

$$\sum_\beta \int E^k \left| \langle \Phi | H - E | \chi_{\beta,\varepsilon_\beta} \rangle \right|^2 dE \approx \sum_q (E_q^{2h1p})^k \left| \langle \Phi | H - E | \Psi_q^{2h1p} \rangle \right|^2 \tag{19}$$

Using the techniques of moment theory, one can recover the correct value of the decay width (10) from the pseudo-spectrum through a series of consecutive approximations of increasing order, [59]. A particularly efficient realization of the Stieltjes imaging procedure that we use in our work has been described in detail in Ref. [66].

C. Selected Applications of the Fano-ADC Method to Interatomic Decay Widths in Clusters

Interatomic decay widths as functions of cluster geometry are not only an essential input for simulations of ICD electron spectra [11, 40], but are also very interesting physical quantities in their own right. The magnitude and the functional form of $\Gamma(R)$ can tell us a lot about physics of interatomic decay. For instance, at large distances, the ICD width can be shown to follow an inverse-power law, in most cases,

$$\Gamma(R) \propto R^{-6} \tag{20}$$

[13, 16, 53], see also [67]. This asymptotic behavior of the decay width can be explained by a physically appealing virtual photon transfer model which represents the decay process as an emission of a virtual photon by the inner-shell-ionized atom followed by its absorption by a neighboring neutral.

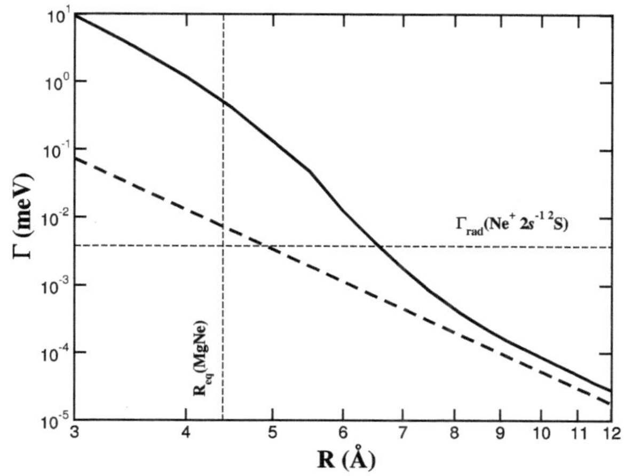

Figure 5: Doubly logarithmic plot of the total ICD width of Ne 2s vacancy in MgNe cluster as a function of internuclear separation. Full line- ADC(2)x result; dashed line – virtual photon transfer prediction [16]. Radiative width of the 2s vacancy in free neon atom [70] and the equilibrium distance of MgNe in the ground state are shown by horizontal and vertical dashed lines, respectively.

The proportionality coefficient in Eq. (20) is thus given by the radiative decay rate of the inner-shell vacancy, total photoionization cross-section of the neighboring atom and a factor that has to do with the decaying state (see Ref. [68] for a general discussion). The virtual photon transfer model neglects the overlap between the atomic orbitals of the atoms participating in the interatomic process and thus its validity around the equilibrium geometry of the neutral cluster was a subject of debate (see, *e.g.* Ref. [69]). Using the Fano-ADC *ab initio* approach of Sections IIIA, IIIB, we have been able to show that the orbital overlap effect leads to a dramatic enhancement of ICD widths in rare gas - alkaline earth diatoms [16, 56] (see Fig. **5**). In rare gas clusters, such as Ne_2, the overlap enhancement of the ICD widths has a moderate character (see Fig. **6**).

The discovery of the overlap effect on the ICD rates led us to ask the question of what kind of chemical environment leads to highest possible ICD rates retaining the clear-cut interatomic nature of the process.

It has been realized quite early [13] that higher ICD rates are favored by environments with the highest possible number of nearest neighbors. In large neon clusters, for example, a "bulk" 2s vacancy would decay faster than the "surface" one [12]. It turns out, though, that even the "bulk" neon ICD rates can be outmatched by interatomic decay in a very interesting group of chemical compounds called endohedral fullerene complexes, *e.g.* in $Ne@C_{60}$ [17]. Indeed, in $(2s^{-1})$ $Ne^+@C_{60}$, the inner valence ionized Ne has as many as 60 nearest neighbors to interact with, which leads to as many as several hundreds of ICD channels (see Fig. **7**).

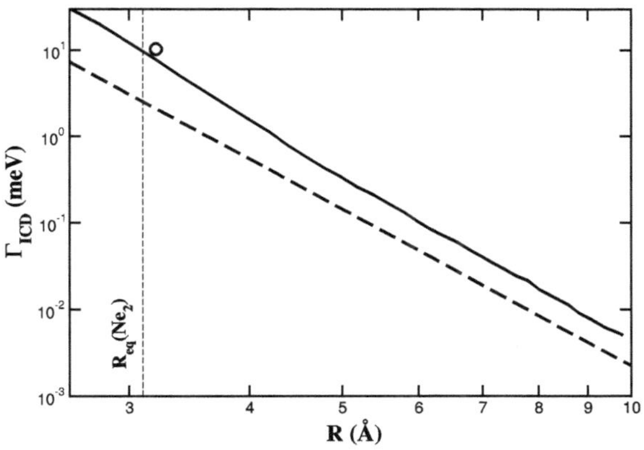

Figure 6: Doubly logarithmic plot of the total ICD decay width of Ne 2s vacancy in Ne_2 cluster as a function of internuclear separation. Full line - ADC(2)x result, dashed line - virtual photon transfer prediction [65], full circle - the result of the single point multi-reference CAP-CI calculation of Ref. [71]. Equilibrium distance of Ne_2 in the ground state is shown by the vertical dashed line.

As a result, the lifetime of ($2s^{-1}$) Ne^+ @C_{60} is only about 2 fs, in fact shorter than Auger lifetime of isolated ls-ionized neon atom [17]. The ultrafast character of ICD is not the only unique feature of interatomic decay in endohedral fullerenes. A detailed consideration of the possible decay pathways reveals that the relatively low multiple ionization energies of the fullerenes give rise to a number of intriguing new processes, such as double ICD (DICD) (being interatomic analog of double Auger decay [72]), double electron transfer mediated decay (DETMD) and a two-step cascade of interatomic decay [17]. A combination of ultra-fast ICD with the rich pattern of the possible electronic decay channels makes endohedral fullerenes particularly attractive for future theoretical and experimental studies. A similar, although less dramatic lifetime reduction has been predicted for atomic inner-shell vacancies embedded in helium droplets [73]. Although helium droplets are usually assumed to be an inert medium used for cooling the encapsulated atoms or molecules, inner-shell ionization triggers weak pairwise ICD interactions between the vacancy and the surrounding He atoms that, once integrated over the droplet volume, produce an appreciable effect on the vacancy lifetime. This effect has been studied for ($2s^{-1}$) Ne+ (Auger-inactive) and ($3p^{-1}$) Ca^+ (Auger-active) vacancy states both for ^4He and for ^3He droplets [73].

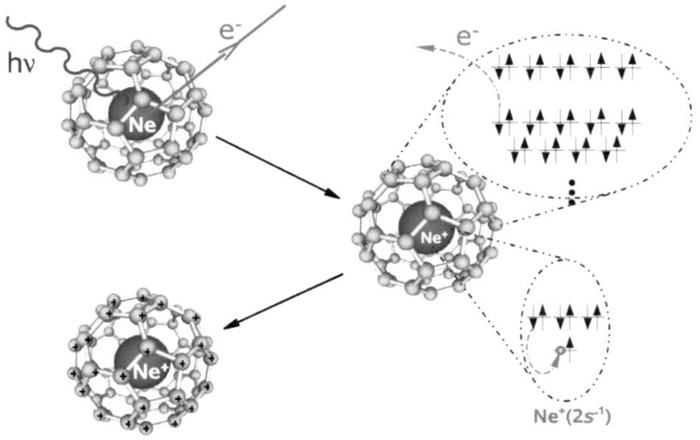

Figure 7: A schematic drawing of the ICD process in ($2s^{-1}$) Ne2@C_{60}: a photon causes inner- valence ionization of the endohedral Ne atom, an outer-valence electron of Ne recombines into the 2s vacancy with the released energy being utilized for fullerene ionization. The predicted ICD lifetime is as short as about 2 fs [17].

IV. INTERATOMIC DECAY IN DOUBLY IONIZED SYSTEMS

As pointed out in Section IB 2, Auger decay of core vacancies often produces highly excited states of the corresponding doubly ionized species which can, in full analogy with the case of single inner-valence vacancy, undergo interatomic (intermolecular) decay of the ICD type in clusters. The last few years have witnessed a significant progress in the experimental investigation of ICD following Auger decay [25, 27, 29-31]. On the theoretical side, the success of the Fano-ADC method, described in Section III, motivated extension of this approach to the description of the interatomic decay of excited doubly ionized states in clusters [74]. Moreover, the time-dependent nuclear dynamics theory of electronic decay has been generalized to the case of decay cascades [32, 33, 75].

A. Dynamics of Auger-ICD Cascades

Let us consider Auger-ICD cascade in neon dimer as an example [32]. The relevant PESs of the Ne_2 computed in [76] are depicted in Fig. **8**.

The whole decay cascade can be described as the following stepwise process:

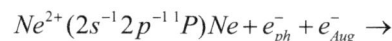

$$Ne_2^+ + \hbar\omega_{exc} \rightarrow Ne_2^+(1s^{-1}) + e_{ph}^- \rightarrow$$

$$Ne^{2+}(2s^{-1}2p^{-1}\,{}^1P)Ne + e_{ph}^- + e_{Aug}^- \rightarrow$$

$$Ne^{2+}(2p^{2}\,{}^{1}D) + Ne^{+}(2p^{-1}) + e^{-}_{ph} + e^{-}_{Aug} + e^{-}_{ICD} \quad \textbf{(21)}$$

Broadband synchrotron radiation with an energy slightly above the K-shell ionization threshold populates the ($1s^{-1}$) Ne^{+} ionic states with emission of the primary photoelectron (e_{ph}). In the next step, the intra-atomic $K — L_1L_{2\ 3}$ (^1P) Auger decay of the ($1s^{-1}$) Ne^{+} core- ionized states populates the one-site dicationic states ($2s^{-1}2p^{-1}\ {}^1P$) $Ne^{2+}Ne$ with emission of an Auger electron (e_{Aug}). The Auger decay lifetime of 2.5 fs is by two orders of magnitude shorter than the typical time scale of the vibrational motion in the core-ionized states ($1s^{-1}$) Ne_2^{+} (300 fs [32]), thus practically excluding possible effect of the nuclear dynamics at the first step of the cascade. Thus, one can assume a vertical (instantaneous) transfer of the nuclear wave packet to the PESs of the final Auger states (see Fig. **8**).

After its creation close to the right turning point of the PES of the ($2s^{-1}2p^{-1}\ {}^1P$) $Ne^{2+}Ne$ state, the intermediate wave packet starts to propagate in the direction of smaller internuclear distances and simultaneously to decay *via* ICD transition into the ($2p^{-2}\ {}^1D$, $2p^{-1}$)Ne^{2}+-Ne^{+} tricationic states with emission of the ICD electron (see Fig. **8**). The potential energy curves for the ($2p^{-2}\ {}^1D$, $2p^{-1}$) Ne^{2+}-Ne+ tricationic states exhibit a typical repulsive character with asymptotic behavior ~ *2/R* a.u. and cross the ($2s^{-1}2p^{-1}\ {}^1P$) $Ne^{2+}Ne$ dicationic states between 2.6 and 2.7 Å. Note, only the uppermost and the lowermost lying repulsive PESs are depicted in Fig. **8**. The typical timescale of the vibrational motion in the ($2s^{-1}2p^{-1}\ {}^1P$) $Ne^{2+}Ne$ states is about 300 fs [32] and is comparable with the lifetime for the electronic decay *via* ICD (100 fs [24]). As shown in [32], the competition between electronic decay and nuclear dynamics in this case influences the computed ICD spectra considerably (see subsection IV D).

The nuclear dynamics accompanying cascade decay (21) can be described in terms of the nuclear wave packets, propagating on the PESs of the initial, intermediate and final states $|\Psi_i(R,t)\rangle, |\Psi_{d_1}(R,\varepsilon_1,t)\rangle, |\Psi_{d_2}(R,\varepsilon_1,t)\rangle$, and $|\Psi_f(R,\varepsilon_1,\varepsilon_2,t)\rangle$, respectively, satisfying the following system of coupled differential equations [32, 33, 75],

$$i\hbar\frac{\partial}{\partial t}|\Psi_i(R,t)\rangle = H_i(R)|\Psi_i(R,t)\rangle \quad i\hbar\frac{\partial}{\partial t}|\Psi_{d_1}(R,t)\rangle = F(R,t)|\Psi_i(R,t)\rangle + \hat{H}_{d_1}(R)|\Psi_{d_1}(R,t)\rangle$$

$$i\hbar\frac{\partial}{\partial t}|\Psi_{d_2}R,\varepsilon_1,t)\rangle = \hat{W}_1(R,\varepsilon_1)|\Psi_{d_1}(R,t)\rangle + (\hat{H}_{d_2}(R)+\varepsilon_1)|\Psi_{d_2}(R,\varepsilon_1,t)\rangle$$

$$i\hbar\frac{\partial}{\partial t}|\Psi_f(R,\varepsilon_1,\varepsilon_2,t)\rangle = \hat{W}_2(R,\varepsilon_2|\Psi_{d_2}(R,\varepsilon_1 t)\rangle$$
$$+(\hat{H}_f(R)+\varepsilon_1+\varepsilon_2)|\Psi^f_m(R,\varepsilon_1,\varepsilon_2,t)\rangle \quad \textbf{(22)}$$

where ε_1 and ε_2 are the energies of the Auger and ICD electrons, respectively; $\hat{H}_{d_1}$ and $\hat{H}_{d_2}$ are the nuclear effective Hamilton operators for the unstable intermediate states D_1 and D_2, respectively, given in the local approximation by expressions like equation (4); $\hat{H}_f$ is the nuclear Hamilton operator for the final state F; *F(R,* t), $W_1(R,t_1)$, and $W_2(R,\ \varepsilon_2)$ are the transition operators between initial state I and intermediate state D_1, between intermediate states D_1 and D_2, and between intermediate state D_2 and final state F, respectively. The nuclear wave packet $|\Psi_f(R,\varepsilon_1,\varepsilon_2,t)\rangle$ on the final state F contains all the necessary information on the spectra of the Auger and ICD electrons. The time evolutions of both spectra, σ_{Aug} (ε_1,t) and σ_{ICD} (ε_2,t), and the spectra, σ_{Aug} (ε_1) and σ_{ICD} (ε_2) themselves, can be computed as follows from [32, 33, 75]:

$$\sigma_{Aug}(\varepsilon_1,t) = \int d\varepsilon_2 \left\|\Psi_f(R,\varepsilon_1,\varepsilon_2,t)\right\|^2, \quad \sigma_{Aug}(\varepsilon_1) = \lim_{t\to\infty}\sigma_{Aug}(\varepsilon_1,t),$$

$$\sigma_{ICD}(\varepsilon_2,t) = \int d\varepsilon_1 \left\|\Psi_f(R,\varepsilon_1,\varepsilon_2,t)\right\|^2, \quad \sigma_{ICD}(\varepsilon_2) = \lim_{t\to\infty}\varpi_{ICD}(\varepsilon_2,t) \quad \textbf{(23)}$$

Equations (22) and (23) suggest that the Auger and ICD spectra can be strongly influenced by the interplay between time duration of the exciting pulse, lifetimes for the Auger and ICD transitions, and time scales for the nuclear dynamics on each decay step.

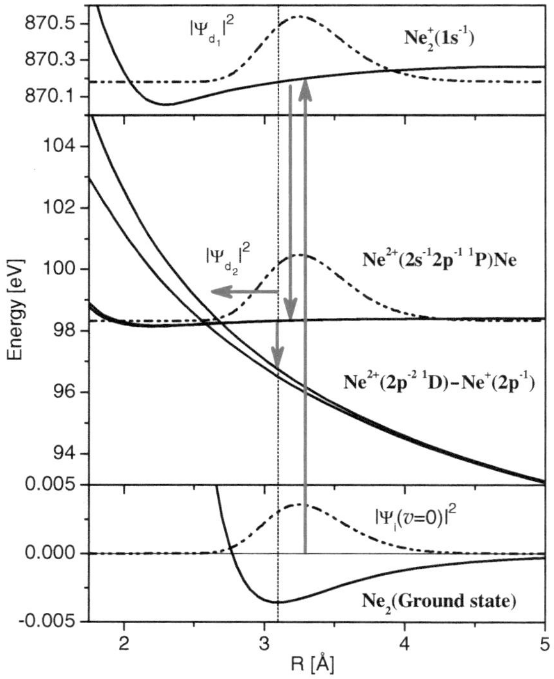

Figure 8: PESs [76] of selected states of Ne_2 involved in the Auger-ICD cascade starting from the core ionization of the Ne dimer and consisting of one-site $K - L_1L_{2,3}\ (^1P)$ Auger transition in the Ne atom followed by ICD transition into the ($2p^{-2}\ ^1D$, $2p^{-1}$)Ne^{2+}-N^+ states. The modulus square of the initial wave packet (dash-dot-dot line) centered around the equilibrium distance R_e = 3.1 Å of the initial state (indicated by a vertical dotted line) is also shown.

B. Fano-ADC Method for Interatomic Decay Widths in (N-2)-Electron Systems

Most computational aspects of the method are fully analogous to the single vacancy case and will not be repeated here. Instead, we will focus only on the few different points. Obviously, the principal difference is that the wave function Ψ_E of Eq.(7) presents now doubly ionized (N — 2-electron) cluster. Therefore, double ionization ADC technique has to be used to construct the bound (Φ) and continuum ($\chi_{\beta,\varepsilon}$)(N — 2)-electron states. Suitable ADC scheme for the *pp*-propagator has been derived by Schirmer and Barth [77]. To describe the bound component (Φ, initial state of the decay) of the wave function Ψ_E within the ADC(n) approach, it is necessary to restrict the physical excitation operators generating the CESs Ψ_J^0 (and, in turn, the intermediate states $\hat{\Psi}_J$) in such a way that open channels of the interatomic decay are not included in the resulting basis.

The intraatomic nature of the Auger decay makes it possible to follow similar strategy as in Section III B, based on the spatial localization properties of the occupied spin-orbitals. Indeed, the initial state of the decay is characterized by two vacancies being localized on a single cluster constituent A, while in the triply ionized final states of Auger-ICD cascade, one or more vacancies reside on another cluster subunit. Therefore, in analogy with Eq. (16) the CESs for the initial state expansion are generated with the restriction that all the holes reside on the subunit A only:

$$\Psi_J^0 = \hat{C}_0^N \qquad (24)$$

$$\{\tilde{C}_J\} \equiv \{c_i c_j, i < j, \phi_{i,j} \in A; c_a^* c_i c_j c_k, i < j < k, \phi_{i,j,k} \in A; ...\}$$

where $\phi_i \in A$ is an occupied spin-orbital of the neutral cluster localized on the subunit A. Note that we are working with two-hole (2h, J=1) three-hole one-particle (3h1p,J = 2).excitation classes. Upon diagonalization of the Hamiltonian constructed in the restricted basis of intermediate states generated from the CESs of Eq. (24), the approximation to the initial state is identified as the eigenstate of the desired symmetry with the largest overlap with the 2h configuration describing the initial two vacancies. The approximate continuum components, $\chi_{\beta,\varepsilon}$, corresponding to the possible final states of the interatomic decay, are obtained in a separate ADC(n) calculation as the eigenstates of the 3h1p character:

$$\chi_{\beta,\varepsilon} \propto \Psi_q^{3h1p} = \sum_i \sum_{|J|=i} Y_{q,J} \tilde{\Psi}_J \quad 1 - \sum_{|J|=2} \left| Y_{q,J} \right|^2 << 1 \tag{25}$$

Once the ADC(n) approximations for the bound and continuum components of the wave function (7) are constructed, one can use the Stieltjes imaging procedure to renormalize the matrix elements computed with the L^2 wave functions, Ψ_q^{3h1p} and interpolate them in energy as necessary for the computation of the decay widths.

C. Auger-ICD Cascades in Rare Gas - Alkaline Earth Clusters

To demonstrate the similarities and differences between ICD in singly and doubly ionized clusters let us investigate the non-radiative decay widths of doubly ionized ($2s^{-1}2p^{-1}$) Ne^{2+}in MgNe diatomic. In the picture provided by the first order of perturbation theory, the decay process proceeds in the same way as ICD of the single Ne 2s vacancy, since the initial Ne 2p vacancy is only a spectator. Higher orders of perturbation theory involve also pathways in which the initial Ne 2p vacancy participates actively, but partial width analysis show that these processes account only for about 5% of the total decay width. The qualitative similarity of the two decay processes is confirmed in Fig. **9**, which shows the total non-radiative decay widths of different symmetries of $MgNe^{2+}(2s^{-1}2p^{-1})$ as functions of internuclear distance R. The decay width of $MgNe^{+}(2s^{-1})$ is shown as the full line for ($2s^{-1}2p^{-1}$) excited state of different spatial symmetries and spin multiplicities. For reference, the full line shows the ICD width of the single Ne 2s vacancy. The equilibrium distance of MgNe in the ground state is shown by vertical dashed line reference. At large distances all the widths follow the $\Gamma(R) \propto R^{-6}$ law, predicted by the virtual photon model for dipole-dipole interatomic decay processes. At smaller distances around the equilibrium geometry the overlap between atomic orbitals of the two involved atoms leads to significant enhancement of ICD widths as in the case of ICD of singly ionized cluster. Note, however, that the overlap enhancement is less pronounced, in particular, in the case of triplet initial states.

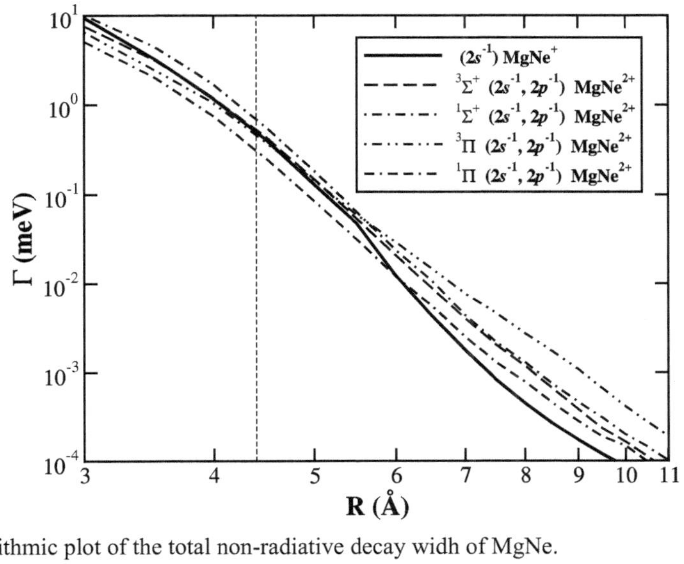

Figure 9: Doubly logarithmic plot of the total non-radiative decay widh of MgNe.

It turns out that, although the spectator Ne 2p vacancy does not affect significantly the qualitative behavior of the interatomic decay rates, it has very strong impact on the magnitude of the widths. For example, at R = 12 Å the decay width of the $^1\Sigma^+$ state of $MgNe^{2+}$ ($2s^{-1}2p^{-1}$) is about 2 times larger than the wdth of the singly ionized cluster and the width of the $^3\Pi$ state is even 3.5 times larger. The change in the magnitude in the presence of additional vacancy is caused mainly by two factors. First, the electron missing in the 2p shell reduces the number of decay pathways, which increases the lifetime of the decaying state. On the other hand, spatial orbitals of multiply ionized atom or molecule are more compact, which increases their Coulomb coupling and, in turn, also the efficiency of the energy transfer dominated pathways of the interatomic decay process. The competition of these two effects explains why, at large distances where the energy-transfer character of the decay prevails, the excited doubly ionized states decay faster. In the orbital overlap dominated region, on the other hand, the compactness of the atomic orbitals leads to reduction of the overlap enhancement, in particular, for the triplet initial states.

To understand more deeply the diverse magnitudes of the total decay widths of various doubly ionized states of different spatial symmetry and spin multiplicity, particular decay pathways and corresponding partial decay widths have to be analyzed. It appears that, for the $^3S^+$ initial state in particular, the vacancy in the 2p shell can impede certain very important decay pathways. For a thorough discussion the reader is referred to Ref. [74].

D. Auger-ICD Cascade in Ne_2

While Auger-ICD cascades in heteronuclear clusters still await their experimental realization, experiments in homonuclear diatomics provide an interesting bulk of data that has been recently analyzed theoretically. The ICD electron spectra after Auger decay of the 1s hole in the Ne dimer was studied in detail in [32]. The time evolutions of the partial ICD spectra were computed by applying the time-dependent theory for the nuclear wave packet dynamics utilizing *ab initio* PECs from [76] shown in Fig. **8**. In the calculations [32], an instantaneous Auger decay of the ($1s^{-1}$) Ne_2^+ states has been assumed in the first step of the cascade, since the lifetime of the $K — LL$ Auger decay in the Ne atom is by two orders of magnitude shorter than the typical time for the vibrational motion in the above state (see discussion in subsection IVA). The following model was applied in [32] in order to simulate the ICD transition rates. First, the R^{-6} analytical behavior reflecting the dipole- dipole nature of the interaction governing the ICD at large distances [16, 71] was assumed. Second, the absolute value of the total ICD rate was equalized to 8.2 meV as computed in [24] at R = 3.2 A (equivalent to the lifetime of 80 fs). Third, the partial decay rates for all ICD channels were taken to be equal, and each partial decay channel has been assumed energetically closed on the left side of the corresponding curve crossing point.

The total ICD spectrum obtained as the sum of the partial spectra over all possible initial and final ICD states assuming their statistical population *via* the $K\text{-}L_1L_{2,3}(^1P)$ Auger decay is compared in Fig. **10a** with the experimental one [30]. In order to illustrate the impact of the underlying nuclear dynamics, the model spectra obtained as a 'mirror-reflection' of the norm of the initial wave packet $|\Psi_i\ (v = 0)\ |^2$ on the final repulsive ICD potential energy curve is also shown in Fig. **10a** by dashed curve. This model supposes instantaneous ICD transition leaving no time for the nuclear dynamics. Comparing the solid and dashed curves in Fig. **10a**, one can see that the nuclear dynamics considerably influences the ICD spectrum. Indeed, the asymmetry in the shape of the total ICD spectrum including the shoulder on its low electron energy side as well as the shift of its maximum to lower energies compared to the vertical electronic transition are due to the nuclear dynamics accompanying the electronic decay.

More recently, the impact of the recoil by the fast Auger electron on the internal nuclear dynamics was observed in [78] by analyzing the partial coincidence ICD spectra following the ls-ionization of the Ne dimer. The interpretation of the recoil effect is based on the theory given in Ref. [79]. The ICD electron spectrum provides a time averaged 'mirror-reflection' image of the nuclear wave packet propagating on the initial PES onto the final repulsive PES [32]. The energy difference between the initial and the final ICD transition states (the energy of the emitted ICD electron) decreases when going to smaller internuclear distances becoming zero at the corresponding crossing point (see Fig. **8**). Obviously, the energy of the emitted ICD electron is closely related to the internuclear geometry at the decay time. During the intra-atomic K-LL Auger decay the fast Auger electron is emitted from the atom where the initially created core-hole was localized, thus, imparting recoil momentum onto this atom which finally becomes $Ne^{2}+$. After the ICD transition has taken place, the Ne+ and $Ne^{2}+$ ions repeal each other strongly, and dissociation along the internuclear axis occurs. If the fast Auger electron is emitted in the direction of the detection of the Ne^{2+} fragment, the recoil momentum is in the direction of Ne^{+} and tends to compress the Ne^{2+}/ Ne system. In the opposite case, if the fast Auger electron is emitted in the direction of the detection of the Ne^{+} fragment, the recoil momentum tends to stretch the dimer. The PESs of the Ne^{2+} Ne initial ICD states, which are the final states of the K-LL Auger decay, are very shallow [76] and have dissociation energy of about 250 meV. The

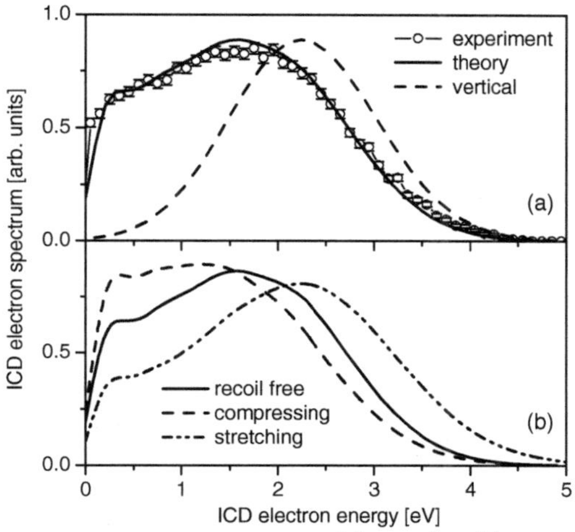

Figure 10: Total ICD electron spectra for the decay of the ($2p^{-2}$ ^1D, $2p^{-1}$)Ne^{2+}-Ne states into the manifold of the ($2p^{-2}$ ^1D, $2p^{-1}$)Ne^{2+}-Ne^{+}) states, after Auger decay of the ($1s^{-1}$) Ne^{+} state. Panel (a): Comparison between the theory [32] and experiment [30]. Panel (b): Computational results [79] for different recoil modes induced by the fast Auger electron.

Auger electron kinetic energy loss due to recoil can be estimated as $\Delta E = E_e \dfrac{m_e}{M_{Ne}} \sim 22$ meV. As can be seen in Fig. **8**, the ICD processes following the emission of a fast electron provide a sensitive tool to study the recoil effect.

It was illustrated [78, 79] that in the compressing mode, the wave packet starts to propagate on the PES of the initial ICD state inwards with an initial velocity provided by the recoil. As a result, the dimer spends more time at smaller internuclear distances, resulting in a preferable emission of the low energy ICD electrons. In the stretching mode, the wave packet starts to propagate outwards with an initial velocity opposite to that in the compressing mode. It turns back and only then moves inwards. As a result, the dimer spends more time at larger internuclear distances, and high energy ICD electrons are preferably emitted. As seen from Fig. **10b** taken from Ref. [79], the total ICD spectrum computed for the compressing mode (dashed curve) is considerably enhanced on the low electron energy side, and in the case of the stretching mode (dash-dot-dotted curve) it is enhanced on the high energy side (to be compared with the total ICD spectrum for the vibrational recoil free mode depicted by a solid curve). A notable difference between the ICD spectra measured for the compressing and stretching modes has been unambiguously observed experimentally [80].

E. Collective ICD

So far, we have considered interatomic decay of doubly ionized states in which one of the holes was a spectator to a very good approximation. It would be interesting to ask, thus, whether a decay process is possible in which both holes will be active. Very recently, it has been predicted that such decay is in fact possible [80]. For this, it turns out, one needs an inner-shell vacancy that is not energetic enough to decay by either Auger or ICD mechanisms. Such are, for example, ns^{-1} states of Ar^{+}, Kr^{+} and Xe^{+}, either isolated or in an environment of other rare gas atoms [81]. Consider two such vacancy states, say, in a Kr cluster or, a bit more generally, in a mixed Kr/X cluster, where X is another atom or molecule. Neither of the vacancies can decay by electron emission, because the energy provided by the 4p $\rightarrow$ 4s recombination is not sufficient for 4p ionization of either Kr^{+} (as needed for Auger decay) or a neutral Kr (as needed for ICD). However, if two 4s-ionized kryptons recombine simultaneously, the released energy would be enough to ionize X:

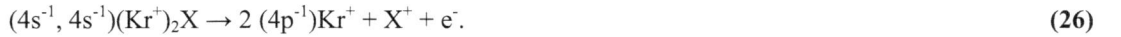

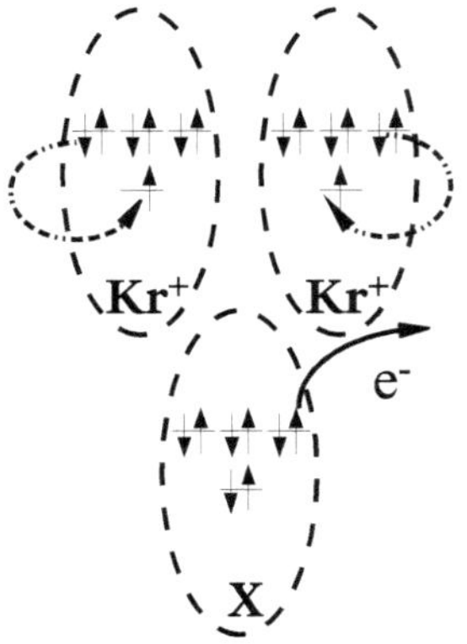

Figure 11: Schematic representation of collective interatomic decay of two inner-shell vacancies: $(4s^{-1}, 4s^{-1})(Kr^+)_2X \rightarrow 2\ (4p^{-1})Kr^+ + X^+ + e^-$.E. Collective ICD.

$$(4s^{-1}, 4s^{-1})(Kr^+)_2X \rightarrow 2\ (4p^{-1})Kr^+ + X^+ + e^-. \quad \textbf{(26)}$$

A schematic representation of such a collective decay process that we call collective ICD (CICD) is given in Fig. **3**. Simple energy considerations imply that, in general, CICD occurs without facing a competition from the ICD if $1.5 < (E_{iv} - E_c)/E_{ion} < 2$, where E_{iv} is the inner valence ionization energy of the given species, E_c is the energy of Coulombic repulsion between two singly ionized atoms or molecules (typically 3-4 eV at the equilibrium distances of neutral clusters) and E_{ion} is the single ionization energy. Looking at the ionization energies of non-metal hydrides [81], for example, one can easily make sure that the collective decay of the type of (26) can occur unhindered by other non-radiative electronic processes in multiply IV-ionized clusters of HCl, HBr, HI, H_2S, H_2Se, H_2Te and PH_3. Notably, the same conclusion holds for multiply IV-ionized clusters of small hydrocarbon molecules (*e.g.* of C_2H_6, C_2H_4, C_2H_2 for gerade IV-orbital ionization).

The collective decay of Eq. (26) is mediated by electronic correlation between three atoms or molecules. Of course, the question arises: how realistic is the proposed decay mechanism? Indeed, if the collective decay is significantly slower than the radiative decay of an isolated atom, it would be suppressed by photon emission. Another possible obstacle for the realization of the new decay mechanism could be the dissociative nuclear dynamics in the decaying doubly ionized cluster.

Feasibility of CICD in Kr_2Ar system has been considered quantitatively in Ref. [80] by nuclear dynamics simulations using Fano-ADC CICD width (see Fig. **12**). It has been concluded that despite the dissociation of the decaying system, CICD yields of up to 65% are possible. The new collective effect enriches the palette of interatomic decay processes discussed in this review.

V. INTERATOMIC DECAY IN INNER-SHELL-EXCITED SYSTEMS

ADC(n) schemes for the excited states of many-electron systems can be described within the ISR approach analogously to how it was done in Section III B for the singly ionized states. The main difference between the ADC(n) approximations for N-electron and (N- 1)-electron systems is the necessary inclusion of the N-electron ground state into the ISR orthonormalization procedure (see Section IIIB), where it plays the role of zeroth excitation class. ADC(1) scheme represents the excited N-electron states in the basis of one hole one particle (1h1p) intermediate states and can be shown to be related to the well-known random phase approximation (RPA) [82]. N-electron ADC(2) uses second-order perturbation theory for the correlated ground state and expands the excited states in 1h1p and 2h2p excitation classes. ADC(2) treats the 1h1p-1h1p and 1h1p-2h2p couplings in second and first order respectively and neglects the coupling between different 2h2p intermediate states. The extended ADC(2) scheme, or ADC(2)x, takes into account the 2h2p-2h2p interactions to first order.

The details of the ISR-ADC schemes for the calculation of the excited states can be found in Ref. [83] and references therein.

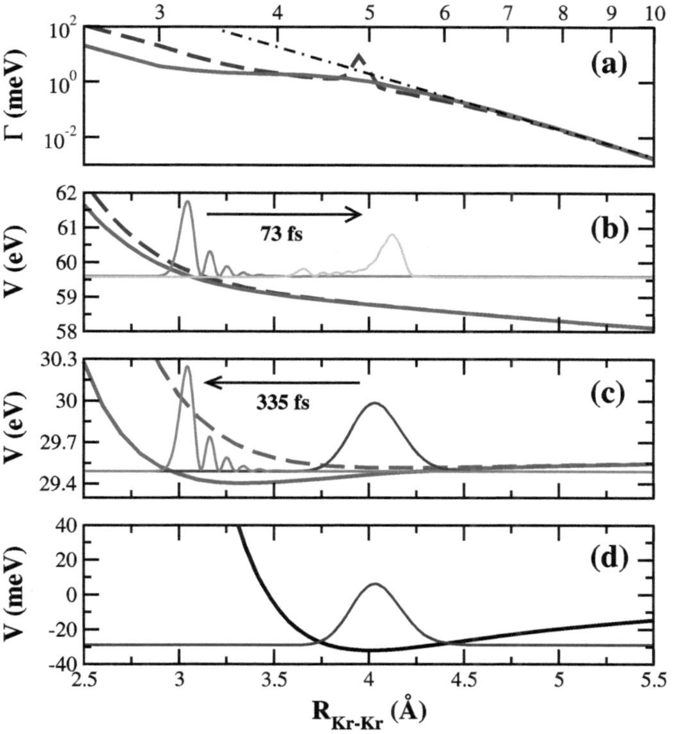

Figure 12: (a) - collective decay widths of ($4s^{-1}$, $4s^{-1}$) $(Kr+)_2Ar$ as functions of $R_{Kr\text{-}Kr}$ at $R^{eq}_{Kr_2-Ar}$ = 3.3Å. Solid line - triplet, dashed line - singlet, dashed-dotted line - $1/R^{10}$ fit (please, note the doubly logarithmic scale). (b) - PESs of ($4s^{-1}$,$4s^{-1}$) $(Kr^+)_2Ar$ states, cut at $R^{eq}_{Kr_2-Ar}$ = 3.3Å. Solid line - triplet state, dashed line - singlet state. Shown also is the vibrational wavepacket promoted from the ($4s\sigma_u^{-1}$) state of $(Kr_2)^+Ar$ [see panel (c)] at the moment of closest approach of the two kryptons, t_{min} = 335 fs, and 73 fs later. Direction of motion of the wavepacket is shown by an arrow. (c) - PESs of ($4s\sigma_{g,u}^{-1}$) $(Kr_2)^+Ar$ states, cut at R_{Kr_2-Ar} = 3.3Å. Solid line - ungerade state, dashed line - gerade state. Shown also is the vibrational wavepacket promoted from the ground state [see panel (d)] at t = 0 (right) and at t_{min} (left). Direction of motion of the wavepacket is shown by an arrow. (d) - PES of the neutral $(Kr_2)Ar$, cut at $R^{eq}_{Kr_2-Ar}$ = 3.3Å. Shown also is the vibrational ground state.

As it was emphasized in Section IB 1, RICD is always accompanied by intraatomic autoionization. Practically, this means that any computational scheme for the decay widths of inner-valence-excited states in clusters must be capable of taking the intraatomic autoionization into account. Let us concentrate for the moment on the problem of computation of autoionization widths of isolated atoms within the Fano-ADC-Stieltjes method and construct the appropriate selection scheme for the 1h1p and 2h2p ADC intermediate states. Suppose, we are interested to compute the decay rates of excited noble gas atoms (Ne, Ar, and Kr), where the excited states belong to the $ms^{-1}np$ Rydberg series (m=2, n>3 for Ne; m=3, n>4 for Ar; m=4, n>5 for Kr) [84]. We see that, due to energy conservation, no singly excited configuration with a hole on the ms orbital or a deeper shell can represent an open decay channel. Therefore, all such configurations should be used in constructing the bound-like state Φ [see Eq. (7)]. Similar considerations can be applied to the doubly excited configurations with one or both of the holes residing on the ms or deeper lying orbitals. We offer the following mathematical formulation of these conditions. The intermediate states $\Psi_J = \hat{C}_J \Psi_0^N$ with

$$\left\{\hat{C}_J\right\} \in \left\{c_a^* c_k, -\varepsilon_k > -\varepsilon_{ms};\ c_a^* c_b^* c_j c_k, a < b, j < k, -\varepsilon_j - \varepsilon_k > -\varepsilon_{ms}\right\} \qquad \textbf{(27)}$$

represent close decay channels and should be used to expand the bound component ϕ. The quantity q is the HF energy of the i-th orbital, or, according to the Koopmans' theorem, the first-order approximation to the respective ionization potential.

The doubly excited configurations $mp^{-2}ab$ with both holes on the outer-valence mp orbital and *a*, *b* standing for virtual orbitals of appropriate symmetry do not appear among the states in Eq. (27). In order to decide whether to include them in the expansion of Φ or not we need to answer the following question.

Is a spectator resonant Auger decay possible in the systems under study?

If the answer is positive, then excited singly ionized species with two holes in an outer-shell are allowed final states and, therefore the configurations $mp^{-2}ab$ cannot be used to expand Φ. Since spectator resonant Auger decay is forbidden in Ne, Ar, and Kr, we can include the configurations $mp^{-2}ab$ among those in Eq. (27). These intermediate states are used to set up an ADC(n) matrix, which is diagonalized and one of its eigenvectors is identified as the bound-like component Φ.

Let us now consider which configurations contribute to the continuum-like part. It is obvious that any singly excited configuration with a hole on an orbital higher than ms represents a valid open channel. Mathematically this condition can be written as $-1 < -t_{ms}$. The allowed final states are of predominantly singly excited character, which might suggest that only singly excited configurations should be used in constructing them. However, $mp^{-2}ab$ states are necessary to describe the electronic correlation in the final state. Thus, they should be included in the expansion of the final states. Since the formulation of the Fano method, which we use, does not call for the orthogonality between the bound and continuum parts, we proceed as follows. The singly excited configurations obeying $-1 < -t_{ms}$ together with $mp^{-2}ab$ states are used to set up another ADC(n) matrix. It is diagonalized and a number of its eigenvectors having the largest weight of singly excited configurations are taken to be the continuum states $\chi_{p,\varepsilon}$. Using this procedure, we improve the description of the final states while avoiding the inclusion of spurious open channels.

We applied the selection scheme described above to compute the decay rates of the excited inner-valence states in Ne, Ar, and Kr atoms by Fano-ADC-Stieltjes method [84]. For all three atoms, we calculated the decay widths Γ of the first three inner-valence-excited-states, *i.e.* $2s^{-1}np$ (n=3,4,5,.) for Ne, $3s^{-1}np$ (n=4,5,6) states for Ar and $4s^{-1}np$ (n=5,6,7) for Kr. The results are shown in Tables **1-3**.

Table 1: Experimental [85] and theoretical decay widths Γ for the autoionizing $2s^{-1}np$ (n=3,4,5) states of Ne. The results of time-dependent density functional theory in the local density approximation (TDLDA) are taken from Ref. [86]. R-matrix (RM) theoretical results are taken from Ref. [87].

	n=3	n=4	n=5
Γ_{exp} (meV)	13 (±2)	4.5 (±1.5)	2(±1)
$\Gamma_{ADC(1)}$(meV)	30.48	9.31	5.59
$\Gamma_{ADC}(2)$ (meV)	8.93	2.86	1.72
Γ_{ADC} (2)x (meV)	11.46	3.78	1.94
Γ_{TDLDA} (meV)	13.90	3.86	1.62
Γ_{RM} (meV)	34.9	6.65	2.47

Table 2: Experimental [88] and theoretical decay widths Γ for the autoionizing $3s^{-1}np$ (n=4,5,6) states of Ar. The TDLDA results are taken from Ref. [86].

	n=4	n=5	n=6
Γ_{exp} (meV)	76 (±5)	25 (±7)	16 (±7)
Γ_{ADC} (i) (meV)	50.61	13.52	5.59
Γ_{ADC} (2) (meV)	61.5	18.42	8.05
Γ_{ADC} (2)x (meV)	67.76	25.85	12.14
Γ_{TDLDA} (meV)	183.4	42.8	18.2

The Fano-ADC-Stieltjes autoionization widths for inner-valence-excited states of Ne and Ar (see Tables **1**, **2**) lie within the experimental error for all n's. The results in the case of Kr (Table **3**) exhibit a comparatively large deviation for the lowest Rydberg term and improve greatly toward higher excitation energies. This is explained by the prevalence of electronic correlation effects in Kr which can only partly be taken into account by the ADC(2)x technique [84]. Theoretical results obtained by TDLDA are comparable to those of Fano-ADC(2)x-Stieltjes data in the case of Ne but become worse than those for Ar and Kr. The R-matrix results, unlike the TDLDA and the present method, fail to reproduce the experimental decay width of the lowest Rydberg $2s^{-1}3p$ term in Ne.

The success of the Fano-ADC-Stieltjes method to describe the intraatomic autoionization calls for the generalization of this technique to the case of interatomic decay of inner-valence- excited states, *i.e.* RICD. The lowest-order perturbation theoretical (Wigner- Weisskopf) estimations of the RICD widths were given in Ref. [23]. The much more accurate Fano- ADC-Stieltjes method could not be readily applied to the RICD problem, however, due to a computational difficulty of the following nature. The standard formulation of the Stieltjes imaging algorithm that we employ [66] requires a full diagonalization of the Hamiltonian matrix which becomes impractical for the excitation ADC(2) and ADC(2)x Hamiltonians, even for diatomic systems, such as Ne_2, once they are represented in high-quality Gaussian basis sets. This drawback of the Stieltjes imaging technique was realized very early on [90] and a few of suggestions as to how it can be overcome exist in the literature [90-92]. Very recently, we have proposed a general method for Stieltjes imaging application to large dimension problems. Our method [93] is based on applying the Stieltjes imaging procedure to the Block-Lanczos pseudospectra [94] instead of the full spectra of the Hamiltonian. The physically sound choice of the initial guess (initial block) for the iterative block-Lanczos procedure can lead to a dramatic reduction of effort required in the combined Lanczos-Stieltjes technique. For example, a converged Lanczos-Stieltjes calculation of the total photoionization cross-section of benzene molecule required a diagonalization of a matrix of the order of 5000 by 5000, while the full ADC(2) Hamiltonian dimension used in the calculation was of the order of 10^6 by 10^6 [93].

Table 3: Experimental [89] and theoretical decay widths Γ for the autoionizing $4s^{-1}np$ (n=5,6,7) states of Kr. The TDLDA results are taken from Ref. [86].

	n=5	n=6	n=7
Γ_{exp} (meV)	22.8 (±0.8)	13.2 (±0.5)	7.8 (±0.6)
Γ_{ADC} (1) (meV)	34.26	9.4	3.97
Γ_{ADC} (2) (meV)	42.37	12.13	5.1
Γ_{ADC} (2)x (meV)	54.69	16.14	7.16
Γ_{TDLDA} (meV)	130.4	27.8	11.6

RICD width of ($2s^{-1}3p$) NeMg obtained by Stieltjes imaging of the Block-Lanczos spectrum of the excitation Fano-ADC(2)x Hamiltonian obtained by the application of the moment theory procedure to the Block-Lanczos pseudospectrum is shown in Fig. **13**. One can see that the discrepancy between the Fano-ADC and the lowest-order Wigner-Weisskopf result can be dramatic, especially for the participator process. Indeed, the Fano-ADC(2)x calculations predict much lower preference for the sRICD (which is still the preferred interatomic decay mode) and an order of magnitude suppression of the spectator process relative to the intraatomic autoionization. Apparently, the interatomic decay will be of much higher importance for the clusters with multiple interatomic decay channels, such as NeAr. Fano-ADC calculations on such systems are now in progress. Future incorporation of the Fano-ADC widths of the RICD process into the nuclear dynamics calculation is expected to yield realistic RICD electron spectra as well as the probability of the cluster disintegration following the RICD process.

VI. OUTLOOK

The preceding sections outlined a remarkable activity, both theoretical and experimental, aiming at in-depth study of interatomic decay processes. A manifold of new physical processes have been observed and still more new phenomena have been predicted theoretically. ICD and the related decay mechanisms has been demonstrated experimentally or predicted theoretically to occur in a wide variety of chemical species: from compact diatomics to such large systems as endohedral fullerenes and helium droplets, from extremely loosely bound helium dimers to hydrogen-bonded water clusters. The first demonstration-of-principle experiment has been performed showing the potential of ICD electron spectroscopy as an analytical technique for the study of interfaces [18]. All these developments clearly point at the study of interatomic decay in clusters as at an emerging field of research. Key to theoretical progress in this new field is the ability to obtain reliable estimations of the rates of the various interatomic decay processes. Here, the Fano-ADC approach described in present paper seems to be the method of choice. While established fairly well for singly and doubly ionized systems, the Fano-ADC technique is yet to be generalized to the case of triply ionized states. Furthermore, it would be desirable to extend the existing Fano-ADC approaches for singly ionized states to the ADC(3) and ADC(4) levels of the *ab initio* theory. Indeed, going to the ADC(3) level of approximation of the many-electron states will help to describe the decay of inner-valence-ionized states more accurately than it is done on the currently implemented ADC(2)x level. This could be very important for the vacancy states of elements beyond the second row of periodic table (*e.g.* $3s^{-1}$ Ar^{+}). ADC(4) level of theory would provide a quantitative description of double decay processes (*e.g.*, DICD) as well as highly accurate results for the decay of core-ionized states. As plausible applications of the envisaged *ab initio* theory, one could cite, for example, the rich pattern of interatomic decay processes in endohedral fullerenes. Of particular interest is the question of the relative time scale of the various single and double decay processes as well as the possible interrelation between ultrafast character of ICD in these systems and the fullerene plasmon.

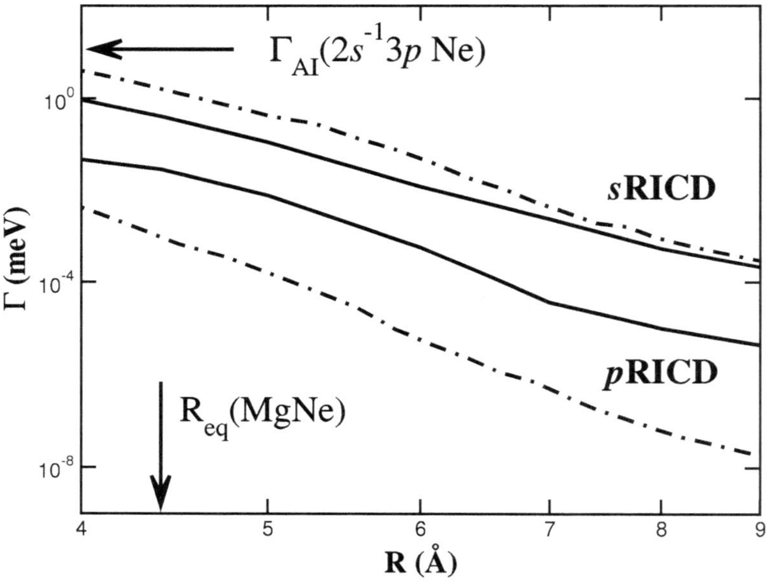

Figure 13: RICD widths of the $(2s^{-1}3p_z)^1\Sigma$ state of MgNe. Solid lines - Fano-ADC(2)x results, dashed-dotted lines - Wigner-Weisskopf results of Ref. [23]. The upper pair of curves show the sRICD widths, while the lower pair of curves represent the pRICD ones. Equilibrium distance of the neutral MgNe cluster and the Fano-ADC(2)x prediction for the $(2s^{-1}3p)$ Ne autoionization are shown by arrows.

While the experiment has so far targeted rare gas clusters, a solid theoretical evidence for interatomic decay in other systems, such as hydrogen bonded clusters and endohedral fullerenes, represents an excellent motivation for bringing some chemical diversity into the experimental studies. First steps along this direction have been taken by Hergenhahn and co-workers who have been able to measure ICD in inner-valence-ionized water clusters [95], Dorner, Jahnke and co-workers who performed coincidence measurements on ICD in water dimer [96] and by Aziz *et al.* who have identified an ICD process involving a ls-ionized OH^- ion and a water molecule in NaOH solution [97]. Relevance of interatomic decay processes for water and water solutions naturally leads to the question of the relevance of ICD for biochemical environment, *e.g.* in the processes leading to radiation damage [98]. At present, this direction is only very little explored and certainly has a high potential for future studies.

A separate chapter in the experimental study of electronic decay has been opened with the advent of new high-frequency radiation sources: attosecond lasers [99] and free electron lasers (FELs) reaching the X-ray regime [100]. Attosecond lasers operating in the XUV domain provide a unique opportunity to study the electronic decay processes in time domain using the so-called streaking probe [101]. The moderate photon energies needed to initiate interatomic decay, together with the characteristic time scales of 1-100 fs make interatomic decay processes natural candidates for the application of the attosecond pump-probe techniques. The first step towards the time-dependent theoretical description of electronic dynamics in the course of interatomic decay has been already taken [102, 103].

In the present chapter, we have outlined the major successes and challenges of the study of interatomic decay with the particular emphasis on the development of *ab initio* theoretical methodology. It is our hope that more theoreticians and experimentalists will enter the new fascinating world of interatomic decay and contribute to turning the new basic concepts into a powerful spectroscopic tool.

ACKNOWLEDGEMENTS

The authors would like to thank Jochen Schirmer for many fruitful discussions in the course of the work described in the present review. V. A. would like to acknowledge financial support of the EPSRC through the Career Acceleration Fellowship (Grant EP/H003657/1). P. K. acknowledges financial support from the Grant GACR 202/09/0786 of the Czech Science Foundation. Ph. V. D. would like to acknowledge the European Commission for the Marie Curie fellowship (PIIF-GA-2008-219224). S. S. would like to acknowledge the Japanese Society for the Promotion of Science (JSPS).. Financial support of Deutsche Forschungsgemeinschaft is gratefully acknowledged. The research leading to these results has received funding from the European Research Council under the European Community's Seventh Framework Programme (FP7/2007-2013) / ERS Advanced Investigator Grant №227597.

DISCLOSURE

"Interatomic Electronic Decay Processes In Clusters": Part of information included in this chapter has been previously published in Journal of Electron Spectroscopy and Related Phenomena Volume 183, Issues 1-3, January 2011, Pages 36-47.

REFERENCES

[1] Auger P. Sur l'effet photoelectrique compose. J de Phys 1925; 6: 205-8.

[2] Burhop EHS, Asaad WN. The Auger Effect. Adv At Mol Phys 1972; 8: 163-8; Mehlhorn W, Auger electron spectroscopy. In: Crasemann Eds. Atomic Inner-Shell Physics. New York and London: Plenum Press 1985.

[3] Turner NH, Schreifels JA. Surface Analysis: X-ray photoelectron spectroscopy and auger electron spectroscopy. Anal Chem 2000; 72: 99R.

[4] Wagner CD, Joshi A. The auger parameter, its utility and advantages: a review. J Elec Spectr Rel Phenom 1988; 47: 283-313.

[5] Cederbaum LS, Zobeley J, Tarantelli F. Giant intermolecular decay and fragmentation of clusters. Phys Rev Lett 1997; 79(24): 4778-81.

[6] Santra R, Zobeley J, Cederbaum LS, Moiseyev N. Interatomic Coulombic decay in van der waals clusters and impact of nuclear motion. Phys Rev Lett 2000; 85(21): 4490-3.

[7] Deleuze MS, Francois JP, Kryachko ES. The Fate of dicationic states in molecular clusters of benzene and related compounds. J Am Chem Soc 2005; 127(48): 16824-34.

[8] Marburger S, Kugeler O, Hergenhahn U, Möller T. Experimental evidence for interatomic Coulombic decay in Ne clusters. Phys Rev Lett 2003; 90(20): 203401.

[9] Jahnke T, Czasch A, Schoffler MS, *et al.* Experimental observation of interatomic Coulombic decay in Neon dimers. Phys Rev Lett 2004; 93(16): 163401.

[10] Dörner R, Mergel V, Jagutzki O, *et al.* Cold target recoil ion momentum spectroscopy: a momentum microscope to view atomic collision dynamics. Phys Rep 2000; 330(2-3): 95-192.

[11] Scheit S, Averbukh V, Meyer HD, *et al.* On the interatomic Coulombic decay in the Ne dimer. J Chem Phys 2004; 121(17): 8393-8.

[12] Öhrwall G, Tchaplygine M, Lundwall, *et al.* Interatomic Coulombic decay in free neon clusters: large lifetime differences between surface and bulk. Phys Rev Lett 2004; 93(17): 173401.

[13] Santra R, Zobeley J, Cederbaum LS. Electronic decay of valence holes in clusters and condensed matter. Phys Rev B 2001; 64(24): 245104.

[14] Vaval N, Cederbaum LS. *Ab initio* lifetimes in the interatomic Coulombic decay of neon clusters computed with propagators. J Chem Phys 2007; 126(16): 164110.

[15] Müller IB, Cederbaum LS. Electronic decay following ionization of aqueous Li^{+} microsolvation clusters. J Chem Phys 2005; 122(9): 094305.

[16] Averbukh V, Müller IB, Cederbaum LS. Mechanism of interatomic Coulombic decay in clusters. Phys Rev Lett 2004; 93(26): 263002.

[17] Averbukh V, Cederbaum LS. Interatomic electronic decay in endohedral fullerenes. Phys Rev Lett 2006; 96(5): 053401.

[18] Barth S, Marburger S, Joshi S, *et al.* Interface identification by non-local autoionization transitions. Phys Chem Chem Phys 2006; 8(27): 3218-22.

[19] Barth S, Joshi S, Marburger S, *et al.* The efficiency of Interatomic Coulombic Decay in Ne clusters. J Chem Phys 2005; 122(24): 241102.

[20] Aoto T, Ito K, Hikosaka Y, Penent Y, *et al.* Properties of resonant interatomic Coulombic decay in Ne dimers. Phys Rev Lett 2006; 97(24): 243401.

[21] Eberhardt W, Kalkoffen G, Kunz C. Measurement of the Auger decay after resonance excitation of Xe 4d and Kr 3d resonance lines. Phys Rev Lett 1978; 41(22): 1569; Brown GC, Chen MH, Crasemann B, Ice GE. Observation of the Auger Resonant Raman Effect, Phys Rev Lett 1980; 45(24): 1937-40.

[22] Gel'mukhanov F, Ågren H. Resonant X-ray Raman scattering. Phys Rep 1999; 312(3): 87-330.

[23] Gokhberg K, Averbukh V, Cederbaum LS. Interatomic decay of inner-valence-excited states in clusters. J Chem Phys 2006; 124(14): 144315.

[24] Santra R, Cederbaum LS. Coulombic Energy Transfer and Triple Ionization in Clusters. Phys Rev Lett 2003; 90(15): 153401.

[25] Morishita Y, Liu XJ, Saito N, *et al.* Experimental evidence of interatomic Coulombic decay from the Auger final states in Argon dimers. Phys Rev Lett 2006; 96(24): 243402.

[26] Liu X J, Saito N, Fukuzawa H, *et al.* Evidence of sequential interatomic decay in argon trimers obtained by electron-triple ion coincidence spectroscopy. J Phys B: A Mol Opt Phys 2007; 40(1): F1.

[27] Morishita Y, Saito N, Suzuki H, *et al.* Evidence of interatomic Coulombic decay in ArKr after Ar 2p Auger decay. J Phys B 2008; 41(2): 025101.

[28] Stoychev S, Kuleff A, Tarantelli F, Cederbaum LS. On the interatomic electronic processes following Auger decay in neon dimer. J Chem Phys 2008; 128(1): 014307.

[29] Kreidi K, Jahnke T, Weber T, *et al.* Localization of inner-shell photoelectron emission and interatomic Coulombic decay in Ne. J Phys B: At.Mol.Opt.Phys. 2008; 41(10): 101002.

[30] Kreidi K, Jahnke T, Weber T, *et al.* Relaxation processes following 1s photoionization and Auger decay in Ne_2. Phys Rev A 2008; 78(4): 043422.

[31] Yagishita A, Stener M, Decleva P, *et al.* Decay channel dependence of the photoelectron angular distributions in core-level ionization of Ne Dimers. Phys Rev Lett 2008; 101(4): 043004.

[32] Demekhin P, Scheit S, Stoychev D, Cederbaum LS. Dynamics of interatomic Coulombic decay in a Ne dimer following the K — Li $L_{2,3}$ (^1P) Auger transition in the Ne atom. Phys Rev A 2008; 78(4): 043421.

[33] Demekhin P, Chiang YC, Stoychev S, *et al.* Recoil by Auger electrons: theory and application. J Chem Phys 2009; 131(10): 104303.

[34] Ueda K. Private communication.

[35] Hultzsch W, Kronast W, Niehaus A, Ruf MW. Investigation of the spontaneous ionisation mechanisms in slow collisions of He^{+} with Ca and Ba, and of Ne^{+} with Ba. J Phys B: At Mol Opt Phys 1979; 12(11): 1821-42, and references therein.

[36] Zobeley J, Santra R, Cederbaum LS. An efficient combination of computational techniques for investigating electronic resonance states in molecules. J Chem Phys 2001; 115(11): 5076-88.

[37] Hergenhahn U. Private communication.

[38] Cederbaum LS, Tarantelli F. Nuclear dynamics of decaying states: A time-dependent formulation. J Chem Phys 1993; 98(12): 9691-706.

[39] Scheit S, Cederbaum LS, Meyer HD. Time-dependent interplay between electron emission and fragmentation in the interatomic Coulombic decay. J Chem Phys 2003; 118(5): 2092-107.

[40] Scheit S, Averbukh V, Meyer HD, Zobeley J, Cederbaum LS. Interatomic Coulombic decay in a heteroatomic rare gas cluster. J Chem Phys 2006; 124(15): 154305.
[41] Havermeier T, Jahnke T, Kreidi K, *et al.* Interatomic Coulombic decay following photoionization of the helium dimer: observation of vibrational structure. Phys Rev Lett 2010; 104(13): 133401.
[42] Sisourat N, Kryzhevoi NH, Kolorenc P, Scheit S, Jahnke T, Cederbaum LS. Ultralong-range energy transfer by interatomic Coulombic decay in an extreme quantum system. Nature Phys 2010; 6(7): 508-11.
[43] Moiseyev N, Santra R, Zobeley J, Cederbaum LS. Fingerprints of the nodal structure of autoionizing vibrational wave functions in clusters: Interatomic Coulombic decay in Ne dimer. J Chem Phys 2001; 114(17): 7351-60.
[44] Gel'mukhanov FK, Mazalov LN, Kondratenko AV. A theory of vibrational structure in the X-ray spectra of molecules. Chem Phys Lett 1977; 46(1): 133-7.
[45] Scheit S, Cederbaum LS. Coincidence and total photoelectron spectra and their differences induced by internal degrees of freedom. Phys Rev Lett 2006; 96(23): 233001.
[46] Moiseyev N. Quantum theory of resonances: calculating energies, widths and cross-sections by complex scaling. Phys Rep 1998; 302(5): 212-93.
[47] Moiseyev N, Scheit S, Cederbaum LS. Non-Hermitian quantum mechanics: Wave packet propagation on autoionizing potential energy surfaces. J Chem Phys 2004; 121(2): 722-5.
[48] Riss UV, Meyer HD. Calculation of resonace energies and widths using the complex absorbing potential method. J Phys B 1993; 26(23): 4503-35.
[49] Muga JG, Palao JP, Navarro B, Egusquiza IL. Complex absorbing potentials. Phys Rep 2004; 395: 357-426.
[50] Santra R, Cederbaum LS, Meyer HD. Electronic decay of molecular clusters: non-stationary states computed by standard quantum chemistry methods. Chem Phys Lett 1989; 303(3-4): 413-19.
[51] Moiseyev N. Derivations of universal exact complex absorption potentials by the generalized complex coordinate method. J Phys B: At Mol Opt Phys 1998; 31(7): 1431-41; Riss UV, Meyer HD. The transformative complex absorbing potential method: a bridge between complex absorbing potentials and smooth exterior scaling. J Phys B: At Mol Opt Phys 1998; 31(10): 2279-304.
[52] Schirmer J, Cederbaum LS, Walter O. New approach to the one-particle Green's function for finite Fermi systems. Phys Rev A 1983; 28: 1237-59.
[53] Santra R, Cederbaum LS. Non-Hermitian electronic theory and applications to clusters. Phys Rep 2002; 368(1): 1-117.
[54] Trofimov AB, Schirmer J. Molecular ionization energies and ground- and ionic-state properties using a non-Dyson electron propagator approach. J Chem Phys 2005; 123(14): 144115.
[55] Müller IB. PhD thesis. Universität Heidelberg 2006, in German. Hennig H. Diploma thesis. Universitat Heidelberg 2004; in German.
[56] Averbukh V, Cederbaum LS. *Ab initio* calculation of interatomic decay rates by a combination of the Fano ansatz, Green's-function methods, and the Stieltjes imaging technique. J Chem Phys 2005; 123(20): 204107.
[57] Fano U. Effects of Configuration Interaction on Intensities and Phase Shifts. Phys Rev 1961; 124(6): 1866-78.
[58] Schirmer J. Closed-form intermediate representation of many-body propagators and resolvent matrices. Phys Rev A 1991; 43(9): 4647-59; Mertins F, Schirmer J. Algebraic propagator approaches and intermediate-state representations. I. The biorthogonal and unitary coupled-cluster methods. Phys Rev A 1996; 53(4): 2140-52; Trofimov AB, Schirmer J. Molecular ionization energies and ground- and ionic-state properties using a non-Dyson electron propagator approach. J Chem Phys 2005; 123(14): 144115.
[59] Langhoff PW. Stieltjes-Tchebycheff moment theory approach in molecular photoionization studies. In: Rescigno T, McKoy V, Schneider B. Eds. Electron-Molecule and Photon-Molecule Collisions. New York: Plenum 1979; p 183; Hazi AU. Stieltjes moment theory technique for calculating resonance widths. In: Rescigno T, McKoy V, Schneider B. Eds. Electron-Molecule and Photon-Molecule Collisions. New York: Plenum 1979; p 281.
[60] Howat G, Åberg T, Goscinski O. Relaxation and final-state channel mixing in the Auger effect. J Phys B: At.Mol.Opt.Phys.1978; 11(9): 1575-88.
[61] Åberg T, Howat In: Handbuch der Physik, Mehlhorn W, Ed. Vol 31. Berlin: Springer 1982.
[62] Szabo A, Ostlund AS. Modern quantum chemistry: introduction to advanced electronic structure theory. New York: Dover 1996.
[63] Schirmer J, Trofimov AB, Stelter G. A non-Dyson third-order approximation scheme for the electron propagator. J Chem Phys 1998; 109(12): 4734-44.
[64] Davidson E. The iterative calculation of a few of the lowest eigenvalues and corresponding eigenvectors of large real-symmetric matrices. J Comp Phys 1975; 17: 87-94.

[65] Averbukh V, Cederbaum LS. Calculation of interatomic decay widths of vacancy states delocalized due to inversion symmetry. J Chem Phys 2006; 125(9): 094107.

[66] Müller-Plathe F, Diercksen GHF. Perturbative-polarization-propagator study of the photoionization cross section of the water molecule. Phys Rev A 1989; 40(2): 696-711.

[67] Matthew JAD, Komninos Y. Transition rates for interatomic Auger processes. Surf Sci 1975; 53(1): 716-25.

[68] Gokhberg K, Kopelke S, Kryzhevoi NV, Kolorenč P, Cederbaum LS. Dependence of interatomic decay widths on the symmetry of the decaying state: analytical expressions and *ab initio* results. Phys Rev A 2010; 81(1): 013417.

[69] Thomas TD, Miron C, Wiesner K, Morin P, Carroll TX, Sæthre LJ. Anomalous natural linewidth in the 2p photoelectron spectrum of SiF_4. Phys Rev Lett 2002; 89(22): 223001.

[70] Griffin DC, Mitnick DM, Randell RN. Electron-impact excitation of Ne. J Phys B: At Mol Opt Phys 2001; 34(22): 4401-15.

[71] Santra R, Cederbaum LS. An efficient combination of computational techniques for investigating electronic resonance states in molecules. J Chem Phys 2001; 115(15): 6853-61.

[72] Carlson TA, Krause MO. Experimental Evidence for Double Electron Emission in an Auger Process. Phys Rev Lett 1965; 14(11): 390-93; Carlson TA, Krause MO. Measurement of the Electron Energy Spectrum Resulting from a Double Auger Process in Argon. Phys Rev Lett 1966; 17(21):1079-82.

[73] Kryzhevoi NV, Averbukh V, Cederbaum LS. High activity of helium droplets following ionization of systems inside those droplets. Phys Rev B 2007; 76(9): 094513.

[74] Kolorenc P, Averbukh V, Gokhberg K, Cederbaum LS. Ab initio calculation of interatomic decay rates of excited doubly ionized states in clusters. J Chem Phys 2008; 129(24): 244102.

[75] Scheit S. PhD thesis. Heidelberg: Ruprecht-Karls-Universitat 2007.

[76] Stoychev S, Kuleff A, Tarantelli F, Cederbaum LS. On the doubly ionized states of Ar2 and their intra- and interatomic decay to Ar^{+}. J Chem Phys 2008; 129(7): 074307.

[77] Schirmer J, Barth A. Higher-order approximations for the particle-particle propagator. Z Phys A 1984; 317: 267-79.

[78] Kreidi K, Demekhin P, Jahnke T, *et al.* Photo- and Auger-electron recoil induced dynamics of interatomic Coulombic decay. Phys Rev Lett 2009; 103(3): 033001.

[79] Demekhin P, Scheit S, Cederbaum LS. Recoil by Auger electrons: theory and application. J Chem Phys 2009; 131(16): 164301.

[80] Averbukh V, Kolorenč P. Collective interatomic decay of multiple vacancies in clusters. Phys Rev Lett 2009; 103(18): 183001; ICD in Ar, Kr and Xe dimers can occur only from satellite states, see Lablanquie P *et al.*, J Chem Phys 2007; 127(15): 154-323.

[81] Potts AW, Price WC. Photoelectron spectra and valence shell orbital structures of groups V and VI hydrides. Proc Roy Soc Lond A 1972; 326: 181-97.

[82] Schirmer J, Mertins F. A new approach to the random phase approximation. J Phys B: At Mol Opt Phys 1996; 29(16): 3559-80.

[83] Schirmer J, Trofimov A. Intermediate state representation approach to physical properties of electronically excited molecules. J Chem Phys 2004; 120(11): 11449-64.

[84] Gokhberg K, Averbukh V, Cederbaum LS. Decay rates of inner-valence excitations in noble gas atoms. J Chem Phys 2007; 126(15): 154107.

[85] Codling K, Madden R, Ederer D. Resonances in the Photo-Ionization Continuum of Ne I (20-150 eV). Phys Rev 1967; 155(1): 26-37.

[86] Stener M, Decleva P, Lisini A. Density functional-time-dependent local density approximation calculations of autoionization resonances in noble gases. J Phys B: At Mol Opt Phys 1995; 28(23): 4973-99.

[87] Schulz K, Domke M, Puttner R, *et al.* High-resolution experimental and theoretical study of singly and doubly excited resonances in ground-state photoionization of neon. Phys Rev A 1996; 54(4): 3095-112.

[88] Sorensen S, Åberg T, Tulkki J, Rachlew-Kalne E, Sundstrom G, Kirm M. Argon 3s autoionization resonances. Phys Rev A 1994; 50(2): 1218-30

[89] Ederer D. Cross-section profiles of resonances in the photoionization continuum of Krypton and Xenon (600-400). Phys Rev A 1971; 4(6): 2263-70.

[90] Nesbet RK. Stieltjes imaging method for computation of oscillator-strength distributions for complex atoms. Phys Rev A 1976; 14(3): 1065-81.

[91] Ivanov VV, Luzanov AV. Semiempirical and *ab initio* calculations of the full configuration interaction using iterated Krylov spaces. J Struct Chem 1997; 38: 10-17.

[92] Carravetta V, Luo Y, Ågren H. Accurate photoionization cross sections of diatomic molecules by multi-configuration linear response theory. Chem Phys 1993; 174(1): 141-53; Ågren H, CarravettaV, Jensen HJA, Jorgensen P, Olsen J. Multiconfiguration linear-response approaches to the calculation of absolute photoionization cross sections: HF, H_2O, and Ne. Phys Rev A 1993; 47(5): 3810-23.

[93] Gokhberg K, Vysotskiy V, Cederbaum LS, *et al.* Molecular photoionization cross sections by Stieltjes-Chebyshev moment theory applied to Lanczos pseudospectra. J Chem Phys 2009; 130(6): 064104.

[94] Parlett BN. The Symmetric eigenvalue problem. Englewood Cliffs: Prentice-Hall 1980.

[95] Mucke M, Braune M, Barth S, *et al.* A hitherto unrecognized source of low-energy electrons in water. Nature Phys 2010; 6: 143-6.

[96] Jahnke T, Sann H, Havermeier T, *et al.* Ultrafast energy transfer between water molecules. Nature Phys 2010; 6(2): 139-42.

[97] Aziz EF, Ottoson N, Faubel M, Hertel IV, Winter B. Interaction between liquid water and hydroxide revealed by core-hole de-excitation. Nature 2008; 455(7209): 89-91.

[98] Vendrell O, Stoychev S, Cederbaum LS. Generation of highly damaging H_2O^+ radicals by inner valence shell ionization of water. Phys Chem Chem Phys. 2010; 12(5): 1006-9.

[99] Hentschel M, Kienberger R, Spielmann C, *et al* Attosecond metrology. Nature 2001; 414(5863): 509-13; Sansone G, Benedetti E, Calegari F, *et al.* Isolated Single-Cycle Attosecond Pulses. Science 2006; 314(3): 443-6.

[100] Kornberg MA, Godunov AL, Itza-Ortiz S, *et al.* Interaction of atomic systems with X-ray free-electron lasers. J Synchrotron Rad 2002; 9: 298-303.

[101] Drescher M, Hentschel M, Kienberger R, *et al.* Time-resolved atomic inner-shell spectroscopy. Nature 2002; 419(6909): 803-7.

[102] Kuleff A, Cederbaum LS. Tracing ultrafast interatomic electronic decay processes in real time and space. Phys Rev Lett 2007; 98(8): 083201.

[103] Saalmann U, Rost JM. Ionization of clusters in strong X-ray laser pulses. Phys Rev Lett 2002; 89(14): 143401.

CHAPTER 3

Photoionization Dynamics: Photoemission in the Molecular Frame of Small Molecules Ionized by Linearly and Elliptically Polarized Light

Danielle Dowek[1*] and Robert R. Lucchese[2]

[1]*Institut des Sciences Moléculaires d'Orsay (ISMO UMR 8214 Univ. Paris-Sud and CNRS), Bât. 350, Université Paris-Sud, 91405, Orsay Cedex, France and* [2]*Department of Chemistry, Texas A&M University, College Station, Texas 77843-3255, USA*

Abstract: Current results and perspectives on the dynamics of photoionization of small molecules obtained from the analysis of Molecular Frame (MF) photoemission are reviewed. The MF observables intrinsically couple the ionized molecular orbital and the scattering wave function of the photoelectron in the ionization continuum. The field of molecular photoionization has benefited from the development of experimental methods based on the determination of electron-ion momentum vector correlations using position and time sensitive detectors, as well as from advanced *ab initio* calculations accounting for electronic correlations which influence the different steps of a photoionization reaction. Key features of valence shell and inner shell photoionization processes, involving resonant and non-resonant mechanisms, are illustrated for small molecules of increasing complexity: we emphasize the general formalism which describes molecular frame photoemission for single ionization of linear molecules induced by linearly and elliptically polarized light after one photon absorption, and its extensions to recoil frame photoemission when non-linear molecules or multiphoton processes are considered. The rich information contained in molecular frame photoemission is at the core of a number of recent experiments at the forefront of molecular physics.

I. INTRODUCTION

The study of the photoionization of molecules has seen remarkable developments in the recent years driven by different motivations. On the one hand, the understanding of molecular photoionization dynamics at the most sensitive level has stimulated a number of experimental and theoretical studies with the "ultimate" goal of allowing a complete determination of the dipole matrix elements of the photoionization transition including their magnitude and phases. Dill recognized that this goal can be reached by recording the photoemission directly in the molecular frame of the ionized target and he derived the first general expression for the fixed-molecule photoelectron angular distribution [1]. Indeed, since molecules are non-spherical objects, their interaction with light depends on their orientation relative to the polarization axis, and the electron photoemission takes place in the molecular frame where it is a function of the electric field orientation with respect to the molecular orientation. More than fifteen years after this theoretical statement, following the pioneering experiments of Golovin *et al.* on valence shell ionization of O_2 [2] and Shigemasa *et al.* on K-shell ionization of N_2 [3], molecular frame photoemission has been determined for one-photon photoionization induced by valence and inner valence shell [4-16] or inner shell [17-33] excitation of a number of mainly diatomic or linear molecules, thereby probing fundamental processes such as electronic correlation [11], circular dichroism in molecular frame photoemission [11,16,20,24], quantum interferences [14,15,26] which may induce symmetry breaking in dissociative ionization of symmetric molecules [12,16,27,31,32], core hole localization and entanglement [28,29].

In most PI reactions that have been studied to date, the scattering of the photoelectron by the anisotropic potential of the residual ion core results in a molecular Frame Photoelectron Angular Distribution (MFPAD) which is generally highly anisotropic and richly structured. The determination of the complex dipole matrix elements [8-11,18,21,35-37], which can be written as $I_{lm\mu}^{(M_i,M_f)} = \left\langle \Psi_i^{(M_i)} \middle| \vec{r}\cdot\hat{n}_\mu \middle| \Phi_f^{(M_f)} \psi_{lm}^{(-)} \right\rangle$ for linear molecules, where M_i and M_f are the total orbital angular momentum of the initial and final states about the

***Address correspondence to Danielle Dowek:** Institut des Sciences Moléculaires d'Orsay, Fédération Lumière Matière, Université Paris-Sud, Bld 350, 91405 Orsay Cedex, France; Tel: +33 (0) 6915 7672; E-mail: Danielle.Dowek@u-psud.fr

Gennadi Ogurtsov and Danielle Dowek (Eds)

molecular axis, m and μ the corresponding projections of the momentum of the electron for each (l,m) partial wave of the scattering wave function and the photon [34], requires an understanding of both the ionized molecular orbital in the initial state and the final state continuum photoelectron orbital that has been scattered by the molecular ion potential. The coherent superposition of the partial waves leading to the MFPAD is then also a probe of the ionization continuum which is sensitive to such processes as shape resonances [18,21] and autoionizing states [14,15,30], coupling between electron and nuclear motions [12,16,27], and in some circumstances subsequent Auger decay [31,32].

Two main philosophies prevail for achieving an experimental determination of the molecular frame photoemission in the study of photoionization. (*i*) The target in an assembly of randomly oriented molecules is excited into a dissociative state of the ionization continuum and the MFPAD is derived from the measured correlated velocity vectors of the photoelectron and atomic photofragments detected in coincidence. This method has been mostly used in the last decade in studies of one-photon ionization, taking advantage of the performance of third generation synchrotron radiation facilities (ALS, BESSY, SOLEIL, ELETTRA, MAXlab, Spring-8...) and very efficient angular resolved electron-ion coincidence techniques, currently based on time and/or position sensitive detectors, and has lead to a wealth of experimental results, such as those cited above and references therein. This method is now extended to the investigation of strong field tunneling ionization in conditions where the laser field also induces dissociation of the ionic ground state [35] (*ii*) The molecule is aligned or oriented with respect to the ionizing light polarization axis using static electric fields arrangements or various laser excitation schemes so that photoelectron angular distributions recorded in the laboratory frame provide access to MFPADs. This approach was pioneered about twenty years ago using the rotationally resolved (1+1′) Resonance Enhanced Multiphoton Ionization (REMPI) [36,37] to record MFPADs for ionization of the $NO(A^2\Sigma^+ \ v=0)$ excited state; this benchmark PI reaction has been revisited using a time-energy mapping of photoelectron angular distributions without the need for rotational state resolution [38]. Recently, efficient adiabatic or nonadiabatic alignment of ground state molecules by intense non-resonant laser irradiation has been demonstrated with a high degree of alignment [39,40]. Combined with subsequent ultra-short ionizing pulses in different schemes, such active molecular alignment techniques are also presently used to investigate molecular frame photoemission in one-photon ionization into bound molecular states [41], as well as fundamental aspects of strong field tunnel ionization of molecules [42-45].

New directions of research have also emerged in the study of molecular dynamics taking advantage of the major developments of the laser based light sources: the use of femtosecond lasers, which is now being extended from the visible range to the VUV and X-ray domain by the advent of free electron lasers such as FLASH and XFEL at Hambourg, LCLS at Stanford and SCSS at Spring-8, and the availability of sub-femtosecond light sources such as the attosecond pulse trains associated with high harmonic generation. Such time resolved light sources enable physicists to employ pump and probe schemes and thus to follow the evolution of a photoinduced unimolecular reaction in unprecedented short time-scale and extended intensity regimes, the femtosecond scale being characteristic of fast nuclear motion and the femto-to-attosecond range that of electronic dynamics. In such studies, photoionization of the evolving molecule often serves as a probe of dissociation or rearrangement processes launched by the pump pulse. As in the study of the dynamics of molecular photoionization, recording photoemission in the molecular frame of the evolving molecular system removes the blurring due to the random orientation of the parent molecule and provides the most complete knowledge about the studied chemical reaction driven by nonadiabatic couplings, and characterized by electron localization or charge migration. In this field, the applicability of the momentum spectroscopy based on coincidence detection of the photoelectron and photofragments has lead to MFPADs for pump and probe pioneering studies of dissociation dynamics of *e.g.* NO_2, $(NO)_2$ and CF_3I using visible femtosecond lasers [46-48], whereas molecular frame photoemission for the characterization of time-resolved electronic-vibrational dynamics of CS_2 was studied combining transient laser alignment with 3D electron imaging [49]. Highlights from studies of molecular dynamics induced by intense ultra-short laser pulses and attosecond VUV trains are presented in two other chapters of this book. The interpretation of such time-resolved sophisticated experiments will obviously benefit from the characterization of the underlying ionization reaction.

At this level it is also interesting to point out the link between the one-photon molecular photoionization and the electron recombination step of the interaction of an intense IR laser field with an assembly of aligned molecules, which can be modeled as the inverse of a PI process: the Highest Occupied Molecular Orbital (HOMO) involved in the PI dipole matrix elements is also encoded in the high harmonic emission signal, amplitudes, phases and polarization, induced by the electron recombination [51-53] leading to the so-called molecular tomography. Molecular frame photoemission in the PI of molecules thereby provides valuable knowledge of interest to these closely related fields [54].

Here we present a review of combined experimental and theoretical studies we have pursued in the recent years, dedicated to molecular frame photoemission studied *via* Dissociative Photoionization (DPI) of simple molecules. This research program has been developed along three main closely coordinated lines.

(*i*) The experimental method relied on the vector correlation (VC) method, which consists of measuring for each studied DPI process the correlated velocity vectors or momenta of the photoelectron and photoion fragments detected in coincidence and collected in a 4π solid angle. After a detailed investigation of the characteristics of molecular frame photoemission in valence shell ionization of diatomic molecules induced by linearly and circularly polarized synchrotron radiation partly reviewed earlier [55], recoil frame photoemission for ionization of small polyatomic non-linear molecules has been explored for inner and outer shell excitation induced by one photon absorption on the one hand, and valence and Rydberg state excitation induced by multiphoton absorption, on the other hand.

(*ii*) A unified formalism describing the generalized MFPAD within the dipole approximation has been developed to rationalize the analysis of the measured angular distribution. First established for one-photon simple ionization of a linear molecule, induced by linearly [9], circularly [10] and elliptically polarized light [11], this formalism provides a simple analytical expansion of the $I(\theta_k, \phi_k, \chi)$ MFPAD, where χ refers to the molecular axis orientation relative to the light polarization axis and (θ_k, ϕ_k) the electron emission direction: all the dynamical information of the PI reaction induced by linearly or elliptically polarized light is contained in four or five one dimensional $F_{LN}(\theta_k)$ functions, respectively, the dependence in terms of the polar angle χ and the azimuthal angle ϕ_k being expressed by simple geometrical functions. The $F_{LN}(\theta_k)$ functions are extracted from the analysis of all coincident events recorded with a 4π collection and enable reconstruction of the MFPADs for any orientation of the molecular axis with respect to the light polarization axis, using the statistics from events in all emission directions. This method has been extended in the two directions explored experimentally, with the goal of exploring the conditions of its applicability to more complex photon-molecule systems. Section III includes both the description of PI of molecules of increasing complexity, from diatomic to small polyatomic molecules with examples of C_{2v} or C_{3v} symmetry species, and that of photoionization induced by multi-photon absorption, which is often met in the pump-probe time resolved schemes discussed above. We show how the observables, now defined as Recoil Frame Photoelectron Angular Distributions (RFPADs), result from the convolution of both the photoionization dynamics, and the molecular structure and/or nuclear rearrangement which may be induced after the ionization reaction has occurred, or even during the multiphoton excitation process.

(*iii*) The studies presented also rely on advanced *ab initio* calculations using the multichannel Schwinger configuration interaction method (MCSCI): this method accounts for fundamental multielectron interactions which govern the electron dynamics at different steps of a photoionization reaction. Such calculations have been performed for most of the PI reactions studied experimentally and compared in detail to the experimental observables. Theoretical methods currently used for the calculations of MFPADs are reviewed in section IV.

In this chapter, we briefly recall basic results for diatomic molecules for which the method has been described previously [55], and focus the reported results on the study of recoil frame photoemission for ionization of small polyatomic linear and non-linear molecules explored for inner and outer shell excitation induced by one photon absorption. We also consider valence and Rydberg state excitation induced by multiphoton absorption, which corresponds to the most recent development.

II. ELECTRON-ION VECTOR CORRELATIONs

The common philosophy of the experiments discussed in this chapter relies on the use of velocity spectrometers [56,57] or reaction microscopes [58,59] which give access to the complete MFPAD for each studied PI reaction, taking advantage of the occurrence of dissociative photoionization (DPI). The Vector Correlation (VC) method consists in measuring the nascent velocity vectors of the ionic fragments and the photoelectron e_{ph} produced in each DPI event, therefore providing a full three-dimensional momentum imaging of the correlated photoelectron and photoion. In the investigation of inner shell photoionization dynamics, considering DPI processes does not constitute a limitation, since in most cases the ionic molecular state produced after the photoelectron ejection further decays by Auger electron emission leading to double, triple or multiple ionization of the target, and breakup induced by the Coulomb interaction. The multiple ionization and dissociative secondary decay reactions offer a number of opportunities to get access to combined photoionization and subsequent fragmentation dynamics diagnostics. For valence shell ionization, the method focuses on the study of the dynamics of PI into excited ionic molecular states lying above the first DPI limit and other strategies have to be considered to study molecular frame photoemission into the bound ionic states such as the ground state and/or lowest excited states of the molecular ion.

The VC method is most efficient when the axial recoil approximation is valid, *i.e.* when the fragmentation dynamics is such that the direction of the detection of fragments produced in dissociative ionization is identical to the orientation of the bond that breaks at the time of the ionization process [60,61]. This condition is fulfilled *e.g.* when the PI reaction populates the ionic molecular ion in a repulsive part of its potential energy curve or surface or when predissociation of a bound ionic state occurs on a time scale shorter than molecular rotation. Detailed discussion of examples of non-axial recoil dissociation dynamics and what can be learned in such conditions will be presented in sections VI-XI, based on the formalism presented in section III.

The complete MFPAD characterizing electron emission in the Molecular Frame (MF) for any orientation of the molecule with respect to the Field Frame (FF) is a function of several angles. The simplest example is that of linear molecules for which the $I(\theta_k, \phi_k, \chi)$ MFPAD induced by linearly or circularly polarized light is a function of three angles. For DPI of a linear triatomic molecule breaking into two heavy fragments and a photoelectron the reaction can be written as:

$$\mathrm{ABC} + h\nu(\hat{\mathbf{e}}) \rightarrow \mathrm{ABC}^{+*} + e_{ph} \rightarrow \mathrm{A}^{+} + \mathrm{BC}^{*} + e_{ph} \qquad \textbf{(1)}$$

$$\mathrm{ABC} + h\nu(\hat{\mathbf{e}}) \rightarrow \mathrm{ABC}^{+*} + e_{ph} \rightarrow \mathrm{ABC}^{++*} + e_{ph} + e_{Au} \rightarrow \mathrm{A}^{+} + \mathrm{BC}^{+*} + e_{ph} + e_{Au} \qquad \textbf{(2)}$$

Considering valence shell and inner shell DPI of ABC molecules, respectively, the VC method leads to the ($\mathbf{V}_{A+}$, $\mathbf{V}_e$, $\hat{\mathbf{e}}$) triplet or the redundant ($\mathbf{V}_{A+}$,$\mathbf{V}_{BC+}$,$\mathbf{V}_e$, $\hat{\mathbf{e}}$) quadruplet for each DPI event, where $\hat{\mathbf{e}}$ stands for the light quantization axis which is either **P**, the polarization axis of linearly polarized light, or **k**, the propagation axis of the light expressed in the Field Frame (FF) or Laboratory Frame (LF). In the axial recoil approximation the $\mathbf{V}_{A+}$ emission velocity is aligned with the molecular axis orientation, and the $I(\theta_k, \phi_k, \chi)$ MFPAD is then coincident with the measured complete angular distribution derived from the spatial analysis of the ($\mathbf{V}_{A+}$, $\mathbf{V}_e$, $\hat{\mathbf{e}}$) vector correlation. In the more general case of a non-linear molecule or non-axial recoil conditions the relationship between the measured RFPAD and the MFPAD is more complex.

II.A. Coincidence Momentum Spectroscopy: from One-Photon to Multi-Photon Ionization

Electron-Ion Coincidence Velocity Spectrometer

Electron-ion velocity spectrometers have been developed taking advantage of the performance of time and position-sensitive detectors used in combination with static fields: uniform extraction electric field, combined with electrostatic lenses or magnetic field appropriately set to collect charged particles emitted from the interaction region with large solid angle efficiency, usually 4π [56,58,59]. The first scheme based on electric field extraction is briefly described here. A well localized cold target, which is a condition for getting good velocity resolution, is obtained at the crossing of a supersonic molecular expansion with the focused light beam, at the center of the velocity spectrometer, as shown in Fig. **1**. The three components of the $\mathbf{V}_{A+}$ ion fragment and $\mathbf{V_e}$ photoelectron emission velocity vectors are deduced from the time of flight (TOF) and impact position of the particles on their respective PSD built up of a stack of Multichannel Plates (MCPs) and a delay-line anode (DLD Roentdek [62]). In the VC spectrometer, the signal detected on the front MCP of the electron detector is usually used as a common start for an eight-independent-channel time-to-digital converter (CTNM-TDC [63], encoding resolution 250 ps or 125 ps) and a Time-To-Amplitude Converter (TAC). The TAC is stopped by the electronic signal synchronous with the ionizing light pulse, providing the electron time of flight (TOF_{elec}) with a coding resolution of 12.5 ps on the 50 ns scale. The time signals at the ends of the two delay lines of each detector are encoded as stop signals in the TDC: four channels are dedicated to the electron detector leading to the electron position, and the four others to the ion detector providing the ion position and time of flight (TOF_{ion}). Typical values of TOF_{elec} and TOF_{ion} with the extraction fields used are in the range of a few tens of ns and a few μs, respectively. Measuring the electron TOF in order to derive the V_{ze} velocity component in the direction of the extraction field requires the use of a pulsed light source such that the period of the temporal structure of the beam is about 100 ns or longer.

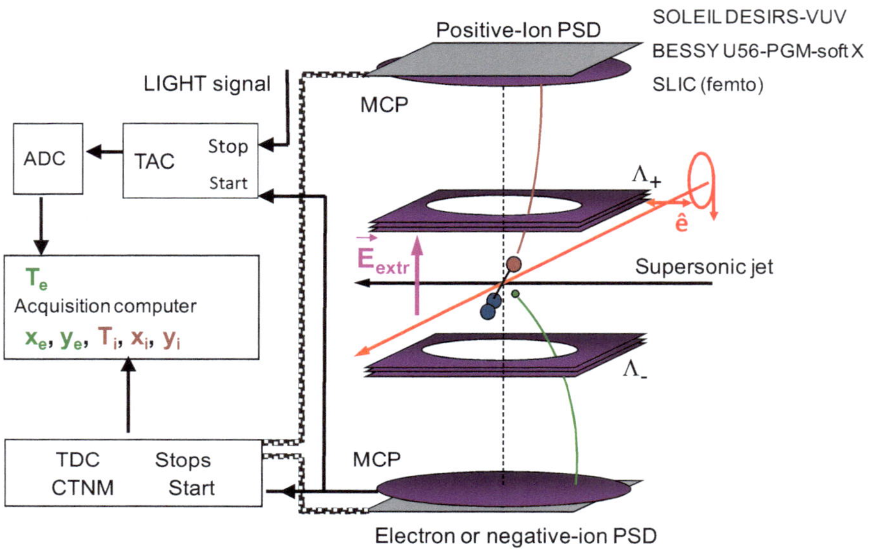

Figure 1: Scheme of the velocity spectrometer and acquisition set-up (see text). PI takes place at the crossing of the molecular beam and the light beam. Positively and negatively charged particles, extracted by a dc uniform electric field combined with electrostatic lenses Λ_+ and Λ_- and collected with a 4π solid angle, are driven to time and position sensitive detectors (PSD) based on delay-line anodes.

Most of the light sources used in the present studies deliver pulses characterized by a time width smaller than a few tens of ps: the time resolution is then limited by the electronic signal shaping contribution, on the order of 100-150 ps in typical devices, corresponding to a spatial resolution of about 200 μm. Recent developments have focused on a significant improvement of the time resolution and have demonstrated a resolution of 18 ps using a femtosecond laser source [64]. The influence of such limitations upon the resolution of the V_x, V_y and V_z components of the velocity vectors, and consequently on the energy and angular resolution, depends on the effective energy of the ionized particles, notably that of the photoelectron. The use of focusing lenses reduces the broadening due to the finite size of the interaction region while preserving a one-to-one $\Delta TOF/V_z$ correspondence, and enables one to work with lower extraction fields ensuring a 4π collection of particles of a given energy. Setting of a magnetic field to control the photoelectron trajectories has extended significantly the tractable electron energy range and enabled the investigation of molecular frame photoemission of Auger electrons [58,59]. When studying inner shell ionization of linear molecules breaking into two singly charged ions, it is of interest to measure the ($\mathbf{V}_{A+}$,$\mathbf{V}_{BC+}$,$\mathbf{V}_e$, ê) quadruplet: the redundancy of the $\mathbf{V}_{A+}$ and $\mathbf{V}_{BC+}$ determination enables an additional correction of the dimension of the interaction region using momentum conservation during the break-up process. There, one uses the large multihit capability per channel of the CTNM, which allows one to register several ions emitted from the same DPI event [24]. The finite deadtime of about 50 ns of the delay-line detectors and read-out electronics may induce a limited loss in the ion collection, which depends on ion fragments mass and energy, and the magnitude of the extraction field, whereas, in the standard conditions, it almost systematically prevents detection of two electrons arising from a single DPI event. Double photoionization studies require special care for the electron detection either using the DL anodes with various geometries and specific electronic coding [65], or relying on *e.g.* multipixel anodes [66]. A discussion of double photoionization is outside the scope of this report.

The extension of the VC method to the study of one-photon PI of non-linear molecules is technically straightforward, however the additional degrees of freedom of the molecule, which may complicate the interpretation of the ($\mathbf{V}_{A+}$, $\mathbf{V}_e$, $\hat{\mathbf{e}}$) and/or ($\mathbf{V}_{A+}$,$\mathbf{V}_{BC+}$,$\mathbf{V}_e$, $\hat{\mathbf{e}}$) observables, require that a careful selection is made of the dissociation channels to be studied. As an example discussed below, the study of L shell $Cl(2p)^{-1}$ ionization of the CH_3Cl methylchloride will be based on the analysis of recoil frame photoemission using so-called complete channels like (CH_3^+, Cl^+) or (H^+, CH_2Cl^+) populated after fragmentation of the CH_3Cl^{++} dication, or (H^+, CH_2^+, Cl^+) produced after double Auger decay, for which energy and momentum conservation can be used to describe the break-up of the cation.

One-Photon Ionization Induced by VUV or Soft X Ray Synchrotron Radiation

Synchrotron radiation that is tunable, polarized, and pulsed with a high repetition rate (1-10 MHz) provides a very appropriate light source for the study of a number of phenomena in PI dynamics using electron-ion coincidence studies. Third generation synchrotron radiation facilities offer a time resolved structure of the light beam *e.g.* single bunch mode with a period of 800 ns, at BESSY and single or eight-bunch mode at SOLEIL, with a period of about 1.1μs or 147 ns, respectively, with the pulse time width being on the order of 50 ps (FWHM). The most complete information is extracted when incident light is circularly or elliptically polarized [11], therefore most of the experiments reported have been performed at BESSY and/or SOLEIL with such polarization states of the light.

We describe the polarization of the light in terms of the four-component (s_0, s_1, s_2, s_3) Stokes vector that determines the total intensity (s_0), the linear intensity (s_1, s_2) and the circularly polarized intensity (s_3) [67]. $(s_1^2 + s_2^2 + s_3^2)^{½}$ and $s_4 = s_0 - (s_1^2 + s_2^2 + s_3^2)^{½}$ represent the intensity of the components of polarized and unpolarized light respectively, and the degree of polarization P is accordingly defined as $P = (s_1^2 + s_2^2 + s_3^2)^{½} / s_0$.

Multi-Photon Ionization Induced by kHz Femtosecond Lasers

The experimental conditions attached to the study of multi-photon PI of linear and non-linear molecules are more challenging for the VC method due to the kHz repetition rate of the femtosecond lasers [68,46,49]. Indeed, performing coincidence measurements requires the occurrence of a maximum of one PI event per pulse, which imposes the constraint that the average number of PI event per pulse must be on the order of a few tenths. In the running conditions, an acquisition rate of a maximum of 50 coincidences/s (0.05 coincident event/pulse) ensures a true-coincidence acquisition mode, whereas current acquisition conditions at the SR source may reach a few times 10^3 c/s with an average number of 0.005 PI event/pulse. Meanwhile, for a better efficiency the timing of the event acquisition may be chosen by using a logical signal synchronous with the 1 kHz laser pulse as the common start for the eight channels of the TDC stopped by the delay line signals. In the reported experiments the light signal also acts as a start for a Time-to-Amplitude Converter (TAC). When turning to XUV incident light, special care has to be given to the filtering of secondary electrons produced by the interaction of stray light with metal surfaces in order to preserve real coincidence conditions.

The experiments reported were carried out using femtosecond lasers of the Saclay Laser-matter Interaction Center (SLIC) facility of the Commissariat à l'Energie Atomique. These are Chirped Pulse Amplification titanium-sapphire (Ti:Sa) laser systems delivering few-mJ pulses at a carrier frequency corresponding to $\lambda_{IR} \sim 800$ nm with a 1 kHz repetition rate. The second or third harmonic pulses used ($\lambda_{air} \sim 400$ nm and ~ 265 nm) are obtained by type I frequency doubling in a BBO crystal. The pulse duration was on the order of 70 fs and 120 fs, respectively.

II.B. Selection of a DPI Process

Once the ($\mathbf{V}_{A+}$, $\mathbf{V}_e$, $\hat{\mathbf{e}}$) and/or ($\mathbf{V}_{A+}$,$\mathbf{V}_{BC+}$,$\mathbf{V}_e$, $\hat{\mathbf{e}}$) vector correlation has been recorded, the path to obtaining the MFPADs requires the selection of a given DPI process. Valence-shell photoionization usually populates a superposition of ionization continua corresponding to different electronic states of the molecular ion and several dissociation limits are observed. Each ionization continuum is characterized by a specific MFPAD for a given excitation energy. We define a DPI process by its reaction pathway in terms of the given molecular state populated after ionization, identified by the photoelectron energy E_e, and the dissociation limit L_D, corresponding to a specific Kinetic Energy Release (KER) of the heavy fragments. Quite often, the different PI processes competing in valence shell ionization cannot be disentangled using solely the photoelectron *or* photoion energy spectrum, and the resolving power of the (E_e, KER) or (E_e, E_{A+}) kinetic energy correlation diagram (KECD) resulting from the analysis of the ($\mathbf{V}_{A+}$, $\mathbf{V}_e$, $\hat{\mathbf{e}}$) correlation is essential for selecting a process and determining the MFPAD characteristics for each ionization continuum. This was demonstrated in several examples of DPI of diatomic molecules [55], and is illustrated here in Fig. **4** of section VI on the example of valence shell PI of the N_2O molecule.

The situation is different for inner shell ionization, since the region of the continuum above a K or L shell ionization threshold is dominated by ionization into the corresponding core hole state, so a single E_e photoelectron energy dominates, except when the ionic state is characterized by fine or hyperfine structure. On the other hand the dissociation channels are more diverse since (*i*) different electronic states of the dication can be populated by simple Auger decay, with their own fragmentation dynamics and dissociation limits, (*ii*) multiple ionization may occur following double or multiple Auger decay populating a number of dissociation channels identified by the mass, the charge state, and the Kinetic Energy (KE) of the atomic and/or molecular ionic fragments.

After selection of a PI reaction as illustrated below, the spatial analysis of the ($\mathbf{V}_{A+}$, $\mathbf{V}_e$, $\hat{\mathbf{e}}$) and/or ($\mathbf{V}_{A+}$,$\mathbf{V}_{BC+}$,$\mathbf{V}_e$, $\hat{\mathbf{e}}$) vector correlation provides the angular distribution which is analyzed in terms of the $I(\theta_k, \phi_k, \chi)$ MFPAD according to the formalism outlined in section III. For linear molecules, this analysis provides the complex dipole matrix elements for the parallel ($\Delta\Lambda$ = 0) and perpendicular ($\Delta\Lambda$ = ± 1) components of the photoionization transition, where Λ is the projection of the total electronic angular momentum onto the molecular axis. The relative phases between the dipole matrix elements characterizing the parallel and perpendicular transitions are obtained from the analysis of the MFPAD for an orientation of the molecular axis which is neither parallel nor perpendicular to the polarization axis of linearly polarized light, such as $\chi = 45°$ or the magic angle, which involves a coherent superposition of both transitions. The most complete information including the sign of these relative phases is obtained when the incident light is circularly or elliptically polarized [10,11] and an extension of this method is presented in section III. In section V we emphasize that the complete and redundant information on a PI reaction provided by the VC method combined with the unified formalism enables the determination of the MFPADs with unknown polarization of the light and, additionally, provides the s_1 and s_2 Stokes parameters characterizing the light polarization state.

If the fragmentation dynamics is fast and no rearrangement occurs, the same information about the ionization dynamics may be obtained from any of the dissociation channels, properly taking into account which channel is recorded. Otherwise, the RFPAD observable results from the convolution of the successive reaction steps.

III. GENERAL FORMALISM FOR MOLECULAR FRAME PHOTOEMISSION

III.A. General Expression of the MFPAD for a Polyatomic Molecule

The differential cross section, $T_{i,f}$, for photoionization in the Molecular Frame (MF) is proportional to the absolute square of the transition matrix elements giving

$$T_{i,f} = \frac{4\pi^2 E}{c} I_{i,f} I^*_{i,f} \tag{3}$$

where E is the photon energy and c is the speed of light. In the dipole approximation, the photoionization matrix element, $I_{i,f}$, can be written in the form,

$$I_{i,f} = \left\langle \Psi_i \middle| B \middle| \Psi^{(-)}_{f,\vec{k}} \right\rangle \tag{4}$$

where Ψ_i is the wave function representing the initial unionized state, $\Psi^{(-)}_{f,\vec{k}}$ is the ionized final state with the continuum electron leaving the system with asymptotic momentum $\hat{k}$, and B is the operator describing the interaction between the field and the electrons in the molecule. For light of arbitrary elliptical polarization, B can be written as

$$B = (\vec{r}\cdot\hat{x}_{\text{FF}})\cos\lambda + (\vec{r}\cdot\hat{y}_{\text{FF}})\sin\lambda\exp(i\delta) \tag{5}$$

where the light is propagating in z direction in the Field Frame (FF), i. e. in the $\hat{z}_{\text{FF}}$ direction, and where $\hat{x}_{\text{FF}}$, $\hat{y}_{\text{FF}}$, and $\hat{z}_{\text{FF}}$ form a mutually orthogonal set of vectors. The parameters λ and δ are related to the Stokes parameters by [67].

$$\begin{aligned} s_0 &= 1 \\ s_1 &= \cos(2\lambda) \\ s_2 &= \sin(2\lambda)\cos(\delta) \\ s_3 &= \sin(2\lambda)\sin(\delta) \end{aligned} \tag{6}$$

In the case of linearly polarized light, equation (5) can be simplified so that equation (4) becomes

$$I^{(\text{LP})}_{i,f} = \left\langle \Psi_i \middle| \vec{r}\cdot\hat{n} \middle| \Psi^{(-)}_{f,\vec{k}} \right\rangle \tag{7}$$

where $\hat{n}$ is the direction of the linear polarization. Defining the spherical tensor operator e_μ by

$$e_\mu = rY_{1,\mu}(\theta_e,\phi_e)\sqrt{\frac{4\pi}{3}} = \begin{cases} z & \text{for } \mu = 0 \\ -\dfrac{x+iy}{\sqrt{2}} & \text{for } \mu = 1 \\ \dfrac{x-iy}{\sqrt{2}} & \text{for } \mu = -1 \end{cases} \tag{8}$$

and writing the partial wave expansion of the ionized state as [69]

$$\Psi^{(-)}_{f,\vec{k}}(\vec{r}) = \sqrt{\frac{2}{\pi}}\sum_{l,m} i^l \Psi^{(-)}_{f,l,m}(\vec{r})Y^*_{lm}(\theta_k,\phi_k) \tag{9}$$

where the angles (θ_k,ϕ_k) define the direction of emission of the photoelectron as illustrated in Fig. **2**, the partial wave dipole matrix elements can then be given by

$$I^{(i,f)}_{lm\mu} = \sqrt{\frac{2}{\pi}} i^l \left\langle \Psi_i \middle| e_\mu \middle| \Psi^{(-)}_{f,l,m}(\vec{r}) \right\rangle \tag{10}$$

$$I_{i,f} = \sum_{l,m,\mu}\sqrt{\frac{4\pi}{3}} I^{(i,f)}_{lm\mu} Y^*_{lm}(\theta_k,\phi_k)Y^*_{1,\mu}(\chi_{\text{LP}},\gamma_{\text{LP}}). \tag{11}$$

In the case of arbitrary polarization, the field can be expressed using $B_\pm$ defined as

$$B_\pm = \cos\lambda \pm i\sin\lambda\exp(i\delta) \tag{12}$$

so that the interaction operator B can be written as

$$B = \frac{1}{\sqrt{2}}\left[B_+ \sum_{\mu=-1}^{1} e_\mu D^{(1)}_{\mu,-1}(\gamma,\chi,\beta) - B_- \sum_{\mu=-1}^{1} e_\mu D^{(1)}_{\mu,1}(\gamma,\chi,\beta) \right] \quad \textbf{(13)}$$

where $D^{(J)}_{M,M'}(\gamma,\chi,\beta)$ are the rotation matrices and (γ,χ,β) are the Euler angles that rotate the molecular frame into the field frame.

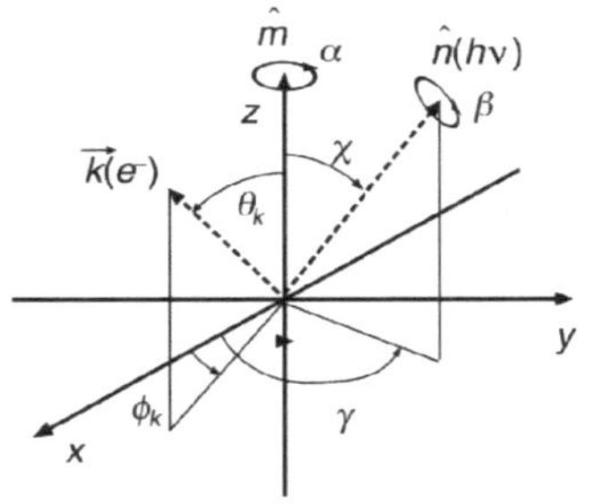

Figure 2: Schematic of the angles and direction used to describe the MFPAD and RFPAD angular distributions. In the schematic, the recoil axis is assumed to be in the *z* direction, $\hat{n}$ is either the direction of light propagation for elliptically polarized light or the direction of polarization for linearly polarized light, and the photoelectron is emitted in direction $\vec{k}$. The angle α represents the rotation of the molecule about the recoil direction and, for elliptically polarized light, the angle β represents the rotation of the $\hat{x}_{\mathrm{FF}}$ direction away from the plane containing the *z* axis and the $\hat{n}$ vector.

Using this notation, the dipole matrix element given in equation (4) can be written in a partial wave expanded form as

$$I_{i,f} = \sum_{l,m,\mu} I^{(i,f)}_{lm\mu} Y^*_{lm}(\theta_k,\phi_k) \frac{1}{\sqrt{2}}\left[B_+ D^{(1)}_{\mu,-1}(\gamma,\chi,\beta) - B_- D^{(1)}_{\mu,1}(\gamma,\chi,\beta) \right] \quad \textbf{(14)}$$

In the case of linearly polarized light or circularly polarized light, equations (11) and (14) can be combined into a single form

$$I^{(\mu_0)}_{i,f} = \sum_{l,m,\mu} I^{(i,f)}_{lm\mu} Y^*_{lm}(\theta_k,\phi_k) D^{(1)}_{\mu,-\mu_0}(\gamma,\chi,\beta) \quad \textbf{(15)}$$

For linearly polarized light in equation (15) we have $\mu_0 = 0$, $\gamma = \gamma_{\mathrm{LP}}$, $\chi = \chi_{\mathrm{LP}}$, and $\beta = 0$. For circularly polarized light $\mu_0 = -s_3$ so that for left-hand circularly polarized light we have $\mu_0 = 1$.

The final differential cross section $T_{i,f}$ as given in equation (3) can be written as

$$T_{i,f} = \frac{4\pi^2 E}{c}\left| \left\langle \Psi_i \middle| B \middle| \Psi^{(-)}_{f,\vec{k}} \right\rangle \right|^2 . \quad \textbf{(16)}$$

$T_{i,f}$ is then, *a priori,* a function of five angles, the electron emission direction in the MF (θ_k,ϕ_k) and the (γ,χ,β) Euler angles that rotate the molecular frame into the field frame (Fig. **2**). The differential cross section can be written in terms of a series of one dimensional functions $F^{(L)}_{N,\nu}(\theta_k)$ times low order functions of the other angles [9]. The $F^{(L)}_{N,\nu}(\theta_k)$ functions contain all the dynamical information and they can be expanded as

$$F_{N,\nu}^{(L)}(\theta_k) = \frac{L!(-1)^{L+N+\frac{|N+\nu|+N+\nu}{2}}}{2\pi} \frac{i^L}{1+\delta_{N,0}} \left[\frac{(L-N)!}{(L+N)!}\right]^{\frac{1}{2}} \times \sum_{L'} \left[\frac{(L'-|N+\nu|)!}{(L'+|N+\nu|)!}\right]^{\frac{1}{2}} P_{L'}^{|N+\nu|}(\cos\theta_k)\, G_{L',L,N,\nu} \quad \textbf{(17)}$$

where the $G_{L',L,N,\nu}$ terms are defined as combinations of products of the partial wave dipole matrix elements:

$$\begin{aligned} G_{L',L,N,\nu} &= \frac{4\pi^2 E}{c} \sum_{l,l',m,\mu} (-1)^{N+m+\mu+1} I_{l,m,\mu}^{(i,f)} \left(I_{l',m+N+\nu,\mu-N}^{(i,f)}\right)^* \\ &\times \left[(2l+1)(2l'+1)\right]^{1/2} \langle 1,1,\mu,N-\mu | L,N\rangle \langle 1,1,1,-1 | L,0\rangle \\ &\times \langle l,l',m,-(m+N+\nu) | L',-(N+\nu)\rangle \langle l,l',0,0 | L',0\rangle \end{aligned} \quad \textbf{(18)}$$

and the $\langle l_1,l_2,m_1,m_2 | L,M\rangle$ are the usual Clebsch-Gordan coefficients. The three indices, *L*, *N*, and *ν*, which define the $F_{N,\nu}^{(L)}(\theta_k)$ functions are constrained by the geometry of the system and the assumptions made in deriving equation (17). Thus in the one-photon dipole approximation, *L* comes from the expansion of the square of the field, so that it satisfies $0 \le L \le 2$, as can be seen in the Clebsch-Gordon coefficients in equation (18). Likewise *N* is constrained by *L* such that $0 \le N \le L$. Note that as the $F_{N,\nu}^{(L)}(\theta_k)$ are defined, only non-negative values of *N* are considered. In principle, the range for L' needed in equation (17) is $0 \le L' \le \infty$. However, in practice, the maximum value of L' needed is limited by the size of the molecular system being considered and by the kinetic energy of the photoelectron. Thus, if the maximum value of *l* needed to describe the asymptotic state of the continuum electron in equation (9) is $l_{\max}$ then the needed range of L' is $0 \le L' \le 2l_{\max}$. Finally, the index *ν* is connected to the amount of asymptotic angular momentum about the $\hat{z}$ axis in the molecular frame. As can be seen in equation (17), the range of *ν* is constrained by the values of L' and *N* leading to the constraint $-L'-2 \le \nu \le L'$. Then the practical limits on *ν* are given by $-2l_{\max} - 2 \le \nu \le 2l_{\max}$. For systems which have a smaller momentum of inertia about the $\hat{z}$ axis compared to the other two axes, the practical limits on the range of *ν* may have a much smaller absolute magnitude.

With the definition of the $F_{N,\nu}^{(L)}$ functions given in equation (17), we obtain the following expression for the $T_{i,f}$ differential cross section in the molecular frame [11,70].

$$\begin{aligned} T_{i,f}(\theta_k,\phi_k,\gamma,\chi,\beta) = \sum_{\nu} \mathrm{Re}\Bigg(& \exp(i\nu\gamma)\Big\{ F_{0,\nu}^{(0)}(\theta_k)\exp\left[i\nu(\phi_k-\gamma)\right] \\ &+ \sum_{N=0}^{1} i s_3 F_{N,\nu}^{(1)}(\theta_k) P_1^N(\cos\chi)\exp\left[i(N+\nu)(\phi_k-\gamma)\right] \\ &+ \sum_{N=0}^{2} F_{N,\nu}^{(2)}(\theta_k)\left[-\frac{1}{2}P_2^N(\cos\chi) + t_1(\beta)Q_N^+(\chi) - it_2(\beta)Q_N^-(\chi)\right] \\ &\times \exp\left[i(N+\nu)(\phi_k-\gamma)\right]\Big\}\Bigg) \end{aligned} \quad \textbf{(19)}$$

where t_1 and t_2 are defined by

$$t_1(\beta) = s_1\cos(2\beta) - s_2\sin(2\beta)$$
$$t_2(\beta) = s_1\sin(2\beta) + s_2\cos(2\beta) \tag{20}$$

and the $Q_N^{\pm}(\chi)$ are defined as

$$Q_N^{\pm}(\chi) = \frac{3}{(2-N)!}\Big\{(-1)^N\left[\cos(\chi/2)\right]^{2+N}\left[\sin(\chi/2)\right]^{2-N} \pm\left[\cos(\chi/2)\right]^{2-N}\left[\sin(\chi/2)\right]^{2+N}\Big\} \tag{21}$$

If the studied molecule is *not chiral*, there is a choice of the molecular frame such that the $F_{N,\nu}^{(L)}$ are real valued functions which leads to further simplification. In particular, equation (19) can then be written as

$$\begin{aligned} T_{i,f}(\theta_k,\phi_k,\gamma,\chi,\beta) = \sum_{\nu}\Bigg(& F_{0,\nu}^{(0)}(\theta_k)\cos(\nu\phi_k) \\ & -\sum_{N=0}^{1} s_3 F_{N,\nu}^{(1)}(\theta_k) P_1^N(\cos\chi)\sin\left[N(\phi_k-\gamma)+\nu\phi_k\right] \\ & +\sum_{N=0}^{2} F_{N,\nu}^{(2)}(\theta_k)\Bigg\{\left[-\frac{1}{2}P_2^N(\cos\chi)+t_1(\beta)Q_N^{+}(\chi)\right] \\ & \times\cos\left[N(\phi_k-\gamma)+\nu\phi_k\right] \\ & +t_2(\beta)Q_N^{-}(\chi)\sin\left[N(\phi_k-\gamma)+\nu\phi_k\right]\Bigg\}\Bigg) \end{aligned} \tag{22}$$

Equation (22) can be simplified for particular choices of the polarization of the light. For *linearly polarized light* we have

$$\begin{aligned} T_{i,f}^{(\mathrm{LP})}(\theta_k,\phi_k,\chi_{\mathrm{LP}},\gamma_{\mathrm{LP}}) = & \sum_{\nu}\sum_{L=0,2}\sum_{N=0}^{L} F_{N,\nu}^{(L)}(\theta_k) P_L^N(\cos\chi_{\mathrm{LP}}) \\ & \times\cos\left[N(\phi_k-\gamma_{\mathrm{LP}})+\nu\phi_k\right] \end{aligned} \tag{23}$$

and for circularly polarized light

$$\begin{aligned} T_{i,f}^{(\mathrm{CP})}(\theta_k,\phi_k,\chi,\gamma) = \sum_{\nu}\Big\{ & F_{0,\nu}^{(0)}(\theta_k)\cos(\nu\phi_k) \\ & \pm\sum_{N=0}^{1} F_{N,\nu}^{(1)}(\theta_k) P_1^N(\cos\chi)\sin\left[N(\phi_k-\gamma)+\nu\phi_k\right] \\ & -\frac{1}{2}\sum_{N=0}^{2} F_{N,\nu}^{(2)}(\theta_k) P_2^N(\cos\chi)\cos\left[N(\phi_k-\gamma)+\nu\phi_k\right]\Big\} \end{aligned} \tag{24}$$

where the upper sign is for left-hand circularly polarized light or helicity of +1.

III.B. General Expression of the RFPADs: From Linear to Small Polyatomic Molecules

For linear molecules we can modify the expressions of the MFPADs in equations (22), (23), and (24) to explicitly include the effects of degeneracy in the initial and final ion states. That is, we must sum over all degenerate final states and average over the degenerate initial states. Here we will not include spin-orbit interactions, in which case the degenerate electronic states can be identified by the angular momentum about the molecular axis. We will write M_i for the electronic angular momentum about the molecular axis of the initial state, which has a degeneracy of g_i, and M_f for the corresponding angular momentum of the ion state. Thus for an electronic Π state, the values for M will be ±1 and g=2. With the sum and average over degenerate states in linear molecules the only non-zero terms remaining are those with $\nu = 0$. Therefore the number of $F_{N,\nu}^{(L)}$ functions describing the MF differential cross section reduces to 4 and 5 for linearly and elliptically polarized light, respectively, and the $G_{L',L,N,0}$ can be written in terms of the partial wave dipole matrix elements as

$$G_{L',L,N,0} = \frac{4\pi^2 E}{g_i c} \sum_{M_i,M_f} \sum_{l,l',m,\mu} (-1)^{N+m+\mu+1} I_{lm\mu}^{(M_i,M_f)} \left[I_{l',m+N,\mu-N}^{(M_i,M_f)} \right]^* \times \left[(2l+1)(2l'+1) \right]^{1/2} \langle 1,1,\mu,N-\mu | L,N \rangle \langle 1,1,1,-1 | L,0 \rangle \times \langle l,l',m,-(m+N) | L',-N \rangle \langle l,l',0,0 | L',0 \rangle \quad \textbf{(25)}$$

The corresponding differential cross sections can then be obtained from equations (22), (23), and (24) by limiting the sums to include only $\nu = 0$ terms, thus leading to explicit equations for the MFPADs as presented previously for linear molecular systems for ionization by linearly [9], circularly [10], and elliptically [11] polarized light, respectively. (Note that in the previously published equations [11] for elliptically polarized light, expressions involving β had the wrong sign in front of this angle). For PI of a linear molecule from an initial neutral state of Σ^+ symmetry to an ionic state of the same Σ^+ symmetry, which is often met in K shell ionization of diatomic molecules such as CO or N_2, the number of independent $F_{N,0}^{(L)}$ functions reduces further to 3 for linearly polarized light [71] and to 4 for and elliptically polarized light. The same is true for a $\Sigma^- \rightarrow \Sigma^-$ transition.

In the axial recoil approximation one assumes that the direction of the detection of fragments produced in dissociative ionization is the same as the orientation of the bond that breaks at the time of the ionization process [60,61]. Each time the axial recoil approximation fails, the measured recoil frame angular distribution (RFPAD) is not the same as the MFPAD. However, we give here some examples where strategies have been developed that probe, to some extent, the predicted MFPADs by measuring the RFPADs, and at the same time achieve original information about the molecular dynamics that causes the breakdown in the axial recoil approximation. For diatomic molecules, the axial recoil approximation will only break down when the molecule has time to rotate before the dissociation occurs. For diatomic molecules and, more generally, linear molecules, if the dissociation can be modeled as a unimolecular process with lifetime τ then one obtains the same functional form of the differential cross section $T_{i,f}$, where the $F_{N,0}^{(L)}$ are modified to include the effects of the rotation which can occur before dissociation leading to [10]

$$F_{N,0}^{(L,\tau)}(\theta_k) = \frac{1}{Q(T)} \sum_{J,J',J''} F_{N,0}^{(L,J,J',J'')}(\theta_k) \exp\left[\frac{-B''J''(J''+1)}{kT} \right] \times \frac{(2J''+1)}{1 + i\frac{\tau B'}{\hbar}[J'(J'+1) - J(J+1)]} \quad \textbf{(26)}$$

where we have assumed a thermal distribution of initial rotational states with temperature T with rotational partition function $Q(T)$, an initial state rotational constant B'' and final state rotational constant B'. The rotational state specific $F_{N,0}^{(L,J,J',J'')}(\theta_k)$ are then defined in terms of rotational state specific $G_{L',L,N,0}^{(J,J',J'')}$, using the relationship given in equation (17) between F and G, which is in turn defined in terms of rotational state specific dipole matrix elements using, in analogy to equation (25),

$$G_{L',L,N,0}^{(J,J',J'')} = \frac{4\pi^2 E}{g_i c} \sum_{M_i,M_f} \sum_{l,l',m,\mu} (-1)^{N+m+\mu+1} I_{l,m,\mu}^{(J,M_f,J'',M_i)} \left[I_{l',m+N,\mu-N}^{(J',M_f,J'',M_i)} \right]^* \\ \times \left[(2l+1)(2l'+1)\right]^{1/2} \langle 1,1,\mu,N-\mu | L,N \rangle \langle 1,1,1,-1 | L,0 \rangle \\ \times \langle l,l',m,-(m+N) | L',-N \rangle \langle l,l',0,0 | L',0 \rangle \quad \textbf{(27)}$$

where the rotational state specific dipole matrix elements are defined as

$$I_{l,m,\mu}^{(J,M_f,J'',M_i)} = \sum_{m',\mu'} I_{l,m',\mu'}^{(M_i,M_f)} (-1)^{m+\mu+M_f-M_i} \frac{2J+1}{2J''+1} \\ \times \sum_K \langle K,J,m+\mu,M_f | J'',m+\mu+M_f \rangle \\ \times \langle K,J,M_i-M_f,M_f | J'',M_i \rangle \langle l,1,m,\mu | K,m+\mu \rangle \\ \times \langle l,1,m',\mu' | K,M_i-M_f \rangle \quad \textbf{(28)}$$

In section VI.A we discuss an example where we used a comparison between the measured and predicted MFPADs to determine the predissociation lifetime of a dissociative ionic state.

For linear polyatomic molecules, a second mechanism for the breakdown of the axial recoil approximation is molecular bending before dissociation. In this case the recoil axis will no longer correspond to the molecular axis at the time of ionization: the situation becomes comparable to the case of DPI of a non-linear molecule, when the recoil direction differs from a symmetry axis, as discussed below. If the angle between the initial direction of the molecular axis and the recoil direction is β_R, then one obtains the $F_{N,0}^{(L,\beta_R)}(\theta_k)$ that include the effects of the bending motion and are given by equation (17) where $G_{L',L,N',0}$ has been replaced by $G_{L',L,N,0}^{(\beta_R)}$ which is defined by [72]

$$G_{L',L,N,0}^{(\beta_R)} = \sum_{J,N'} \langle L',L,N,-N | J,0 \rangle \langle L',L,N',-N' | J,0 \rangle \\ \times P_J^0(\cos\beta_R)(-1)^{N'-N} H_{L',L,N',0} \quad \textbf{(29)}$$

In section VI.B we discuss the possibility of estimating the extent of bending that occurs in the dissociation of an ionic state by comparing the measured and computed RFPADs.

We now turn to non-linear molecules. In the *recoil frame*, to obtain the RFPAD one must average over different orientations of the molecule, which can be obtained by a rotation about the recoil axis, since this orientation cannot be determined experimentally when only two fragments are produced. If the rotation of the molecule about the recoil axis is described by the angle α then equation (19) becomes

$$T_{i,f}(\theta_k,\phi_k,\gamma,\chi,\beta,\alpha) = \sum_\nu \mathrm{Re}\Bigg(\exp\left[i\nu(\gamma-\alpha)\right] \Big\{ F_{0,\nu}^{(0)}(\theta_k) \\ \times \exp\left[i\nu(\phi_k-\gamma)\right] \\ + \sum_{N=0}^{1} i s_3 F_{N,\nu}^{(1)}(\theta_k) P_1^N(\cos\chi) \exp\left[i(N+\nu)(\phi_k-\gamma)\right] \\ + \sum_{N=0}^{2} F_{N,\nu}^{(2)}(\theta_k) \left[-\frac{1}{2} P_2^N(\cos\chi) + t_1(\beta) Q_N^+(\chi) - i t_2(\beta) Q_N^-(\chi) \right] \\ \times \exp\left[i(N+\nu)(\phi_k-\gamma)\right] \Big\} \Bigg) \quad \textbf{(30)}$$

Then when averaging over α using

$$T_{i,f}^{(\mathrm{RF})}(\theta_k,\phi_k,\gamma,\chi,\beta)=\frac{1}{2\pi}\int_0^{2\pi} T_{i,f}(\theta_k,\phi_k,\gamma,\chi,\beta,\alpha)d\alpha \tag{31}$$

all terms with $v \neq 0$ will average to zero leading to the RF differential cross section:

$$\begin{aligned} T_{i,f}^{(\mathrm{RF})}(\theta_k,\phi_k,\gamma,\chi,\beta) &= \mathrm{Re}\left\{ F_{0,0}^{(0)}(\theta_k)+\sum_{N=0}^{1} is_3 F_{N,0}^{(1)}(\theta_k)\right. \\ &\quad \times P_1^N(\cos\chi)\exp\left[iN(\phi_k-\gamma)\right] \\ &\quad +\sum_{N=0}^{2} F_{N,0}^{(2)}(\theta_k)\left[-\frac{1}{2}P_2^N(\cos\chi)+t_1(\beta)Q_N^+(\chi)-it_2(\beta)Q_N^-(\chi)\right] \\ &\quad \left.\times\exp\left[iN(\phi_k-\gamma)\right]\right\} \end{aligned} \tag{32}$$

For molecules that do not have chiral centers and when the recoil axis is in a reflection symmetry plane of the molecule, this equation leads to

$$\begin{aligned} T_{i,f}^{(\mathrm{RF})}(\theta_k,\phi_k,\gamma,\chi,\beta) &= F_{0,0}^{(0)}(\theta_k)-\sum_{N=0}^{1} s_3 F_{N,0}^{(1)}(\theta_k) \\ &\quad \times P_1^N(\cos\chi)\sin\left[N(\phi_k-\gamma)\right] \\ &\quad +\sum_{N=0}^{2} F_{N,0}^{(2)}(\theta_k)\left\{\left[-\frac{1}{2}P_2^N(\cos\chi)+t_1(\beta)Q_N^+(\chi)\right]\cos\left[N(\phi_k-\gamma)\right]\right. \\ &\quad \left.+t_2(\beta)Q_N^-(\chi)\sin\left[N(\phi_k-\gamma)\right]\right\} \end{aligned} \tag{33}$$

which is formally identical to equation (22), and, by extension, to equations (23) and (24) for linearly and circularly polarized light, with the restriction that $v = 0$. The functional form of the RFPAD is therefore identical to that obtained for the MFPAD of a linear molecule [9-11]. This same functional form can be obtained for a more general choice of recoil axis in non-chiral molecules as long as one averages the RFPAD over pairs of axes that are connected by a reflection through a symmetry plane of the molecule.

Additionally, one can distinguish between two cases of RFPADs based on the relationship between the recoil axis and a rotation symmetry axis of the molecule. First, if the recoil axis is in the direction of a symmetry axis of the molecule, then the matrix elements for parallel and perpendicular transitions relative to the recoil axis will be distinct, leading to a closer connection between the MFPAD and RFPAD. This is *e.g.* the case of dissociative ionization of CH_3Cl when the CH_3^+ and Cl^+ fragments recoil along the C-Cl z C_{3v} symmetry axis discussed in section VII.

Second, we consider the case where the recoil direction differs from a symmetry axis. The matrix elements $I_{lm\mu}^{(i,f)}$ are then usually computed in an initial reference frame that is not simply related to the recoil frame, but where the recoil frame is obtained from the initial frame by a rotation through a second set of Euler angles $(\alpha_\mathrm{R},\beta_\mathrm{R},\gamma_\mathrm{R})$. One finds that equation (17) can be used if $G_{L',L,N,v}$ is replaced by $G_{L',L,N,v}^{(\alpha_\mathrm{R},\beta_\mathrm{R},\gamma_\mathrm{R})}$, which is characteristic of the geometry of the dissociation process and is defined by [73].

$$\begin{aligned} G_{L',L,N,v}^{(\alpha_\mathrm{R},\beta_\mathrm{R},\gamma_\mathrm{R})} &= \sum_{J,N',v'} G_{L',L,N',v'}\left(\frac{2J+1}{2L'+1}\right)\langle J,L,v,N|L',N+v\rangle \\ &\quad \times\langle J,L,v',N'|L',N'+v'\rangle\left[D_{v',v}^{(J)}(\alpha_\mathrm{R},\beta_\mathrm{R},\gamma_\mathrm{R})\right]^* \end{aligned} \tag{34}$$

Note that β_R in equation (34) is the same as that used in equation (29).

Dissociative photoionization of the C_{2v} NO_2 molecule into (NO^+ + O + e) features an example where this formalism must be applied, as illustrated in section VIII. In this case, the RFPAD averaging always mixes at least two different transition dipole symmetry directions, i. e. *x* with *y*, *x* with *z*, or *y* with *z*, making it more difficult to extract the underlying MFPAD from an observed RFPAD.

In non-linear molecular systems, the breakdown of the axial recoil approximation may be due to rotation and bending as discussed for linear molecules, however it may also be created by a more global change of geometry of the molecule such as a rearrangement or an isomerization taking place during dissociation. Again it is possible to infer information about the relationship between the orientation of the molecular frame at the time of ionization and the direction of the recoil fragments from the comparison of RFPADs from experiment and theory using equation (34).

When the RFPAD is written in terms of the $F_{N,0}^{(L)}$ functions, as for example for linearly polarized light

$$T_{i,f}^{(\mathrm{LP})} = \sum_{L=0,2} \sum_{N=0}^{L} F_{N,0}^{(L)}(\theta_k) P_L^N(\cos\chi_{\mathrm{LP}}) \cos\left[N(\phi_k - \gamma_{\mathrm{LP}})\right] \tag{35}$$

and assuming that a 4π collection of photoelectrons and recoil ions is achieved experimentally, it is easy to determine the four $F_{N,0}^{(L)}$ functions with a very good accuracy by a fit of the experimental data to the above equation, which is equivalent to a (χ,ϕ) 2D Fourier transform of the measured distribution [9,10]. Alternatively, four pairs of values of χ_{LP} and $(\phi_k - \gamma_{\mathrm{LP}})$ [74] constitute the minimum number of fixed space directions for which measuring the $I(\theta_k)$ distribution gives access to the four $F_{N,0}^{(L)}$ functions. Using a coincidence dissociative ionization experiment to determine the MFPAD with the detection at a few fixed molecular axes orientations relative to the field requires the measurement of the photoelectron at a range of angles relative to the light polarization. Depending on the experimental geometry and detection schemes, it may be advantageous to use instead detection in the Electron Frame (EF). In this case [75]

$$T_{i,f}^{(\mathrm{LP,EF})} = \sum_{L=0,2} \sum_{N=0}^{L} F_{N,0}^{(L,\mathrm{EF})}(\theta_k) P_L^N(\cos\chi_{\mathrm{EF}}) \cos\left[N(\phi_{\mathrm{EF}} - \gamma_{\mathrm{EF}})\right] \tag{36}$$

where θ_k is the same polar angle between the direction of emission of the photoelectron and the recoil axis, χ_{EF} is the angle between the direction of the field polarization and the direction of emission of the photoelectron, ϕ_{EF} is the azimuthal angle of the recoil axis in the electron frame and γ_{EF} is the azimuthal angle of the polarization direction in the electron frame. In this frame, one can determine the Electron frame Photoelectron Angular Distribution (EFPAD) by measuring the photoelectron at a few orientations relative to the field orientation while measuring the full angular range of ionic fragment emission. Once the $F_{N,0}^{(L,\mathrm{EF})}(\theta_k)$ functions have been determined, a simple rotation recovers the Molecular Frame (MF) functions $F_{N,0}^{(L,\mathrm{MF})}(\theta_k)$ [76],

$$\begin{aligned} F_{N,0}^{(L,\mathrm{MF})}(\theta_k) = \sum_{N'} \left[\frac{(L-N)!(L+N')!}{(L+N)!(L-N')!}\right]^{1/2} \frac{(-1)^{N'}}{1+\delta_{N,0}} \\ \times\left[d_{-N,-N'}^{(L)}(\theta_k) + (-1)^N d_{N,-N'}^{(L)}(\theta_k)\right] F_{N',0}^{(L,\mathrm{EF})}(\theta_k) \end{aligned} \tag{37}$$

III.C. RFPADs for Multiphoton Ionization of Small Polyatomic Molecules

Another approach to obtain new information about the photoionization dynamics within the axial recoil approximation for linear and non-linear molecules is to study multi-photon dissociative ionization, which most often involves ionization of electronically excited states. In the weak field limit, the n-photon ionization process can be analyzed in terms of different pathways involving a series of intermediate states leading to ionization. If there is one bound state, ψ_{n-1}, that is the last electronically excited state before ionization that is common to the important pathways, then the total differential cross section can be factored as

$$T_{i,f}^{(n)}(\theta_k,\phi_k,\gamma,\chi,\beta)=T_{i,n-1}^{(n-1)}(\gamma,\chi,\beta)T_{n-1,f}(\theta_k,\phi_k,\gamma,\chi,\beta) \tag{38}$$

where $T_{i,n-1}^{(n-1)}(\gamma,\chi,\beta)$ is the probability for exciting state ψ_{n-1} for the given orientation of the molecule and $T_{n-1,f}(\theta_k,\phi_k,\gamma,\chi,\beta)$ is the one-photon cross section for ionization from state ψ_{n-1} which can be parameterized as discussed above for one-photon ionization. In general that angular dependence of the bound-to-bound intensity can be expanded in terms of its dependence on the orientation of the molecule about the recoil axis as

$$T_{i,f}^{(n-1)}(\gamma,\chi,\beta)=\sum_{\nu'=-2(n-1)}^{2(n-1)} H_{\nu'}^{(n-1,\lambda,\delta)}(\chi,\beta)\exp(-i\nu'\gamma) \tag{39}$$

where the sum over ν' is limited by the number of photons absorbed and λ and δ characterize the elliptical polarization of the light field. This is a generalization of the results for bound-to-bound transitions given by Dixon [77]. Taking the product of this function and the one-photon ionization intensity given in equation (19) in the general case, and performing the average over the orientation about the recoil axis, leads to

$$\begin{aligned}T_{i,f}^{(n,\mathrm{RF})}(\theta_k,\phi_k,\gamma,\chi,\beta)=\sum_{\nu=-2(n-1)}^{2(n-1)} \mathrm{Re}\Bigg[& H_{\nu}^{(n-1,\lambda,\delta)}(\chi,\beta) \\ & \times\Bigg(F_{0,\nu}^{(0)}(\theta_k)\exp\left[i\nu(\phi_k-\gamma)\right] \\ & +\sum_{N=0}^{1} is_3 F_{N,\nu}^{(1)}(\theta_k)P_1^N(\cos\chi)\exp\left[i(N+\nu)(\phi_k-\gamma)\right] \\ & +\sum_{N=0}^{2} F_{N,\nu}^{(2)}(\theta_k)\exp\left[i(N+\nu)(\phi_k-\gamma)\right] \\ & \times\left\{-\frac{1}{2}P_2^N(\cos\chi)+\left[t_1(\beta)Q_N^+(\chi)-it_2(\beta)Q_N^-(\chi)\right]\right\}\Bigg)\Bigg]\end{aligned} \tag{40}$$

Note that the average over the orientation about the recoil axis leads to non-zero terms only when $\nu=\nu'$, thus leading to the single sum over ν as given in equation (40). Thus, if there is sufficient knowledge about the bound-to-bound part of the excitation probability, i. e. $H_{\nu}^{(n-1,\lambda,\delta)}(\chi,\beta)$, then it is possible to extract information about the $F_{N,\nu}^{(L)}$ for non-zero values of ν from experimentally measured recoil-frame multi-photon dissociative photoelectron angular distributions. As discussed above, for linear molecules the only non-zero $F_{N,\nu}^{(L)}$ functions are those with $\nu=0$. Thus for linear molecules, the number of $F_{N,\nu}^{(L)}$ which need to be determined is the same as in the case of one-photon ionization, i. e. four $F_{N,\nu}^{(L)}$ for linearly polarized light and five otherwise. Note that one can show that $F_{0,0}^{(1)}$ is zero for linear molecules.

IV. THEORETICAL METHODS FOR THE STUDY OF MOLECULAR FRAME PHOTOEMISSION

Any method capable of computing photoionization matrix elements [78] can be used to study MFPADs. Likewise any method that can compute electron-molecular ion scattering wave functions can in principle be used to compute photoionization matrix elements. One approach for computing the photoionization matrix elements is to compute separately the wave functions for the initial and final electronic state at a particular geometry. Then the appropriate dipole matrix integral is computed using the explicit functions for the initial and final states. The initial state, Ψ_i, can be treated with traditional Hartree-Fock (HF) or Configuration Interaction (CI) methods appropriate for such a bound state wave function. The final state, $\Psi^{(-)}_{f,\vec{k}}$, must be treated differently since the final state is part of the continuum and the wave function does not go to zero when one of the electrons is far from the molecule. Instead, the appropriate boundary condition for the ionized state has a flux of electrons leaving the system with asymptotic momentum $\vec{k}$. The final state is often expanded in terms of a close-coupling expansion, *i.e.* a finite sum of products of *N*–1 electron target states, Φ_p, times one-electron continuum functions, $\psi^{(-)}_{p,\vec{k}}$. In some formulations, additional purely bound *N* electron terms, $\Psi^{(b)}_q$, are also added to the expansion leading to the form

$$\Psi^{(-)}_{f,\vec{k}} = \sum_{p=1}^{N_c} A\left(\Phi_p \psi^{(-)}_{p,\vec{k}}\right) + \sum_{q=1}^{N_b} \Psi^{(b)}_q \tag{41}$$

where *A* is the antisymmetrization operator. Such an expansion can be characterized by the number of terms included in the expansions and the form of the target states used in the expansion. Thus one can have Single-Channel (SC) and Multi-Channel (MC) expansions. Additionally, if the target states are described using single configuration state functions where the orbitals are the same as in an initial state Hartree-Fock calculation, then this is referred to as a frozen-core Hartree-Fock calculation (FCHF). Alternatively, one can use some type of CI wave function in the target states.

The close-coupling approach has been extensively used to study molecular photoionization within the Multichannel Schwinger Configuration Interaction Method (MCSCI) [79-81]. Two additional methods based on the wave function expansion given in equation (41) that have been extensively applied to electron-molecule collisions but have been only applied to electron-molecular ion collisions in a limited number of cases are the complex Kohn variational method [82], and the R-matrix method [83]. Numerical solutions of the close-coupling equations have also been obtained using methods based on *B* spline basis sets [84].

One of the difficulties with direct wave-function-based approach is that the initial and final states are computed using separate calculations, and a balanced treatment of these two parts of the calculation cannot be uniquely defined. An alternative approach for obtaining photoionization cross sections is to use linear-response theories of electronic structure theory. In general, response theories for the electronic structure of a molecule solve the appropriate equations for the response of the electronic wave function to an oscillating external field. The simplest linear response theory for molecular calculations assumes that the wave function of the unperturbed state is represented by a HF wave function. With this assumption, the resulting equations are usually referred to as the Random Phase Approximation (RPA), or as the Time-Dependent Hartree-Fock (TDHF) approximation [85,86].

Beyond traditional *ab initio* type methods, molecular photoionization can be treated using Density Functional Theory (DFT) formulations. The simplest application of the DFT is to use a form of the one-electron Kohn-Sham (KS) Hamiltonian to describe the one-electron continuum states [87]. A more accurate DFT method for treating molecular photoionization is based on the Time-Dependent Density Functional Theory (TDDFT) method [88]. This approach is the DFT analog of the RPA method discussed above since it is based on the linear response of a HF like wave function to a slowly time varying external electric field.

One of the first computational methods that was used to study molecular photoionization is the Multiple Scattering Method (MSM) [89]. This approach assumes a very simplified electron-molecule interaction potential for which the corresponding scattering equation can be easily solved. This approach has been applied to many photoionization systems with significant success in describing the non-resonant scattering and the various one-electron resonances which occur. The advantage of this approach is that it can be applied to large systems without significant computational effort.

Of the methods for computing molecular photoionization matrix elements mentioned above, most have been used to compute MFPADs. Here we will give a representative publication for each approach. The

MCSCI method has been used many times for such studies. One particularly interesting study was of the valence ionization of NO where the effects of using different close coupling expansion on the MFPAD were considered in some detail [11]. The simpler SC type calculations have also been used to study MFPADs in C 1s ionization of CO_2 [22]. The complex Kohn method has also been used to study the MFPADs in C 1s ionization of CO_2 [90]. The TDDFT method was used to study the full MFPAD of H_2O [33]. The RPA was used to study the inner shell ionization of N_2 [21], and the MSM was used to study the RFPADs of valence ionization of CH_3F and CH_3Cl [7]. The *B* spline method has been used to study PADs in both single and double ionization of H_2 [27,91]. Finally, although the R-matrix method has not yet been applied to MFPAD type calculations, promising results on the total photoionization cross sections of N_2 and NO [92] indicate that this could also be a good approach for computing MFPADs.

Beyond molecular photoionization, similar theoretical techniques can be used to study electron continuum processes encountered in high-field physics. Such processes include high-harmonic generation and rescattering phenomena [93].

V. A COMBINED DETERMINATION OF MOLECULAR FRAME PHOTOEMISSION AND LIGHT POLARIZATION STATE IN DPI OF LINEAR MOLECULES INDUCED BY ELLIPTICALLY POLARIZED LIGHT

The determination of MFPADs in the study of axial recoil dissociative ionization of linear molecules is usually attached to the use of well defined linearly and/or circularly polarized light. In this section, we show that, using the VC method and the formalism reported in section III, MFPADs, as well as the polarization state of the light, can be derived from a PI experiment induced by unknown elliptically polarized light [94], and, as far as MFPADs relative to linear polarization are concerned, with unpolarized light. For an illustration of these properties we consider the simplest case of PI of a diatomic molecule in axial recoil conditions, and we select here K(O)-shell site selective ionization of the CO molecule studied at BESSY (beam line UE 56 PGM1), for a photon excitation energy $h\nu = 550$ eV, leading to the production of two ionic fragments $C^+ + O^+$ after Auger decay. The UE 56 PGM elliptically polarized undulator beamlines at BESSY II were well characterized for selected photon energies [95], using the BESSY soft X-ray polarimeter based on phase retarding transmission-and linear-polarizing reflection multilayers [96]. The general reaction is written as

$$\begin{aligned} &CO(X\,^1\Sigma^+) + h\nu(\hat{\mathbf{e}}) \rightarrow CO^+\{^2\Sigma^+\,(\mathrm{O\ 1s})^{-1}\} + e_{ph} \\ &\rightarrow CO^{++} + e_{ph} + e_{Au} \rightarrow C^+ + O^+ + e_{ph} + e_{Au} \end{aligned} \qquad \mathbf{(42)}$$

The VC method relies on the analysis of the ($\mathbf{V}_{C+}$, $\mathbf{V}_{O+}$, $\mathbf{V}_e$, $\hat{\mathbf{e}}$) quadruplet, which involves the measured emission velocity vectors of the two ionic fragments and the photoelectron e_{ph}. The experiment was performed with both elliptically polarized light of propagation axis **k**, and linearly polarized light **P** where $\hat{\mathbf{e}}$ denotes the light quantization axis, **k** or **P**. The KER distribution of the ionic fragments shows the population of several $\{C^+ + O^+\}$ dissociation limits assigned to selected states of the dication, consistent with previous measurements [97]. We select for the angular analysis the decay channel through the lowest $CO^{++}(^1\Pi$ and $^3\Sigma^+)$ states assigned to the KER structure centered at 7-8 eV, dissociating into the $C^+(^2P)$ + $O^+(^4S)$ ground-state dissociation limit, giving the largest ion fragment anisotropy $\beta_r \approx 0.6$. Using any elliptically polarized light, including unpolarized light, β_r is determined by analyzing the laboratory frame photoion polar angle distribution with respect to the light propagation axis, according to the expression:

$$I(\chi) = I_0\left\{1 - \frac{\beta_r}{2} P_2(\cos\chi)\right\} \qquad \mathbf{(43)}$$

and the general form of the MFPAD $I(\theta_k, \phi_k, \chi, \beta)$ (from equation (19) with $\nu = 0$ and real valued $F_{N,\nu}^{(L)}$) for a linear molecule is written as:

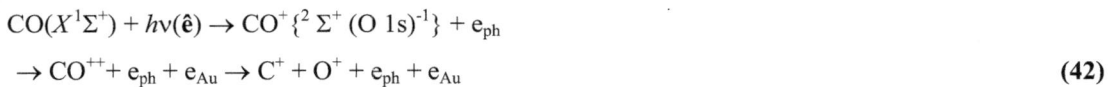

$$\begin{aligned}
I(\theta_k,\phi_k,\chi,\beta) &= F_{00}(\theta_k) \\
&+ F_{20}(\theta_k)\left[-\frac{1}{2}P_2^0(\cos\chi)+t_1(\beta)Q_0^+(\chi)\right] \\
&+ F_{21}(\theta_k)\left\{\left[-\frac{1}{2}P_2^1(\cos\chi)+t_1(\beta)Q_1^+(\chi)\right]\cos(\phi_k)\right. \\
&\left.+t_2(\beta)Q_1^-(\chi)\sin(\phi_k)\right\} \\
&+ F_{22}(\theta_k)\left\{\left[-\frac{1}{2}P_2^2(\cos\chi)+t_1(\beta)Q_2^+(\chi)\right]\cos(2\phi_k)\right. \\
&\left.+t_2(\beta)Q_2^-(\chi)\sin(2\phi_k)\right\} \\
&- s_3F_{11}(\theta_k)P_1^1(\cos\chi)\sin(\phi_k)
\end{aligned} \quad \textbf{(44)}$$

where we keep for convenience the $F_{LN}(\theta_k)$ notation introduced previously [9] in the general expression of the MFPADs for a linear molecule, with the correspondence $F_{LN}(\theta_k) = F_{N,0}^{(L)}(\theta_k)$ using the general notation in equation (17).

We recall that t_1 and t_2 are related to the Stokes parameters s_1 and s_2, as defined in equation (20), by the rotation through the angle β which is the third Euler angle in the rotation matrix defined in section III (related to equation (13)) and the functions $Q_N^{\pm}(\chi)$ are defined in equation (21).

A four angle fit of the $I(\theta_k,\phi_k, \chi, \beta)$ measured distribution based on the functional form given in equation (44) enables one to determine

(*i*) The four $F_{LN}(\theta_k)$ functions (F_{00}, F_{20}, F_{21}, F_{22}) providing the complete set of MFPADs for any orientation of the molecular axis with respect to the polarization axis of linearly polarized light,

(*ii*) The product of the fifth F_{11} function by the Stokes parameter s_3,

(*iii*) The Stokes parameters s_1 and s_2.

This evaluation can be visualized in two steps [94]. Firstly, the $I(\theta_k, \phi_k, \chi, \beta)$ distribution can be reduced to the following function of three angles after integration over the photoion azimuthal angle β:

$$\begin{aligned}
I(\theta_k,\phi_k,\chi) &= F_{00}(\theta_k) - s_3F_{11}(\theta_k)P_1^1(\cos\chi)\sin(\phi_k) \\
&-\frac{1}{2}F_{20}(\theta_k)P_2^0(\cos\chi)-\frac{1}{2}F_{21}(\theta_k)P_2^1(\cos\chi)\cos(\phi_k) \\
&-\frac{1}{2}F_{22}(\theta_k)P_2^2(\cos\chi)\cos(2\,\phi_k)
\end{aligned} \quad \textbf{(45)}$$

This expression is formally identical to the general expression established for PI of a linear molecule by circularly polarized light [10,11], except for the circular dichroism term F_{11} which is here multiplied by s_3. The method developed previously applies then for the extraction of the (F_{00}, F_{20}, F_{21}, F_{22} and $s_3\,F_{11}$) functions.

Knowing the (F_{00}, F_{20}, F_{21}, F_{22}) functions one can plot the MFPADs for any orientation of the molecular axis χ with respect to the polarization axis of linearly polarized light using

$$\begin{aligned}
I\chi(\theta_k,\phi_k) &= F_{00}(\theta_k) + F_{20}(\theta_k)P_2^0(\cos\chi) \\
&+F_{21}(\theta_k)P_2^1(\cos\chi)\cos(\phi_k) \\
&+F_{22}(\theta_k)P_2^2(\cos\chi)\cos(2\phi_k)
\end{aligned} \quad \textbf{(46)}$$

and after determination of s_3 as discussed below, that for a molecular axis oriented relative to the propagation axis of circularly polarized light of helicity +1:

$$\begin{aligned} I\chi(\theta_k,\phi_k) &= F_{00}(\theta_k)+F_{11}(\theta_k)P_1^1(\cos\chi)\sin(\phi_k) \\ &-\frac{1}{2}F_{20}(\theta_k)P_2^0(\cos\chi)-\frac{1}{2}F_{21}(\theta_k)P_2^1(\cos\chi)\cos(\phi_k) \\ &-\frac{1}{2}F_{22}(\theta_k)P_2^2(\cos\chi)\cos(2\phi_k) \end{aligned} \quad \textbf{(47)}$$

Secondly, the ion fragment angular distribution $I(\chi, \gamma)$ obtained from $I(\theta_k, \phi_k, \chi, \beta)$ after integration over the photoelectron emission angles leads to the unambiguous determination of the Stokes parameters s_1 and s_2 and the asymmetry parameter β_r, provided that the latter is non zero:

$$I(\chi,\beta)=C\left\{1+\beta_r\left[-\frac{1}{2}P_2^0(\cos\chi)+\frac{1}{4}t_1(\beta)P_2^2(\cos\chi)\right]\right\} \quad \textbf{(48)}$$

The best accuracy in the extraction of s_1 and s_2 is obtained for processes with large β_r photoion asymmetry parameters. If the degree of polarization of the light P is close to 1, $|s_3|$ is determined as well. Practically, for a given polarization state preparation on the beamline, the sign of the helicity of the light is usually known, therefore the sign of s_3 is known, and the present analysis provides the precise value of the three Stokes parameters at the time the experiment is performed. In this case, the fifth $F_{11}(\theta_k)$ function is therefore also obtained. Otherwise, the determination of the intrinsic sign of s_3 must rely on a complementary information about the sign of F_{11}, either from theory, or from a previous study of the PI reaction. In such cases, the determination of s_3 requires the existence of a non-zero circular dichroism in the molecular frame.

For the process given in equation (42) studied using a single elliptical polarization of the light on the beam line UE 56 PGM1 at $h\nu$ = 550 eV, known to be adjusted for a positive helicity ($s_3 < 0$), the fit of the $I(\chi, \beta)$ distribution according to equation (48) leads to: $\beta_r \approx 0.55 \pm 0.05$, $s_1 \approx 0.75 \pm 0.05$ and $s_2 \approx 0.1 \pm 0.05$, then $s_3 \approx$ -0.66 ± 0.05.

Selected cuts of the MFPADs derived from the experiment performed with the elliptically polarized light are displayed in Fig. **3**(a-d) for the most meaningful orientations of the molecular axis: $\chi = 0°$, 90° and 135° relative to the linear polarization axis **P**, and χ = 90° relative to the propagation axis **k** of circularly polarized light, and the relevant azimuthal angles (ϕ_k = 0° and 180°) and (ϕ_k = 90° and 270°), respectively. They feature the MFPADs for the parallel and perpendicular transitions (a,b), and those illustrating the linear (c) [24,97,98] and circular (d) [20,24,94,98] dichroisms in inner shell PI involving a coherent superposition of parallel and perpendicular transitions, characterized by significant left-right scattering asymmetries in this representation. These MFPADs compare fairly well with those determined experimentally using linearly polarized light at BESSY and the VC method in comparable conditions, shown in Fig. **3** (*e.g.*), and those measured previously using the COLTRIMS technique at $h\nu$ = 553.7 eV [97]. The measured MFPADs are very well described by the MCSCI calculations performed for K(O)-shell PI at the same photoelectron energy, also included in Fig. **3**.

We point out that the method described in this section to determine the light polarization state does not require the use of synchrotron radiation in the few bunch mode, since it relies on the measurement of the photoion distribution only. The polarization state is obtained with good accuracy using electron-ion coincidence detection, even when the electron TOF is not determined. The VC method has recently been compared to optical polarimetry using the VUV polarimeter located on the DESIRS beamline [99,100], leading to a very consistent prediction of the Stokes parameter .

The determination of the Stokes parameters from the $I(\chi,\beta)$ photoion angular distribution is related to previous work based on photoelectron spectroscopy [102 and ref. therein]. These studies dealt with both extraction of s_1 and s_2 for the characterization of linearly polarized light in non-coincidence PES and extraction of the three (s_1, s_2, s_3) Stokes parameters by suitably selected photoelectron-Auger electron coincidence studies [102].

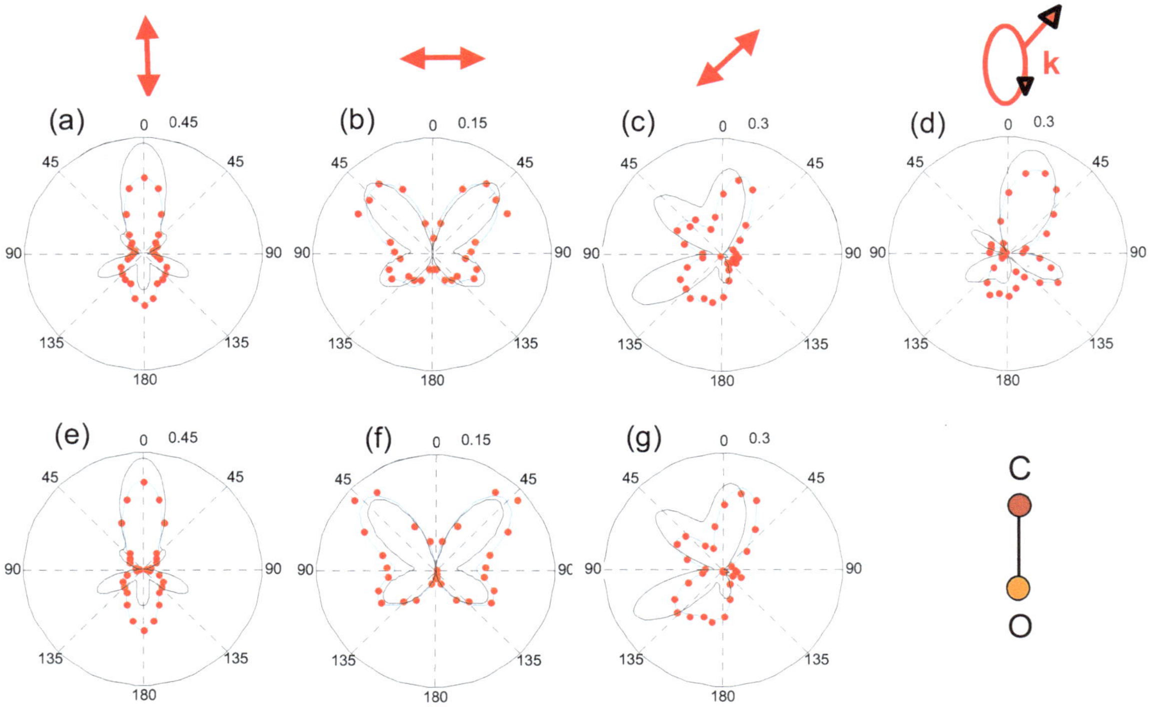

Figure 3: Measured (dots and fit line through the dots) and computed (full line) MFPADs for K(O)-shell PI of CO at hν = 550 eV, for a recoil direction parallel ($\chi = 0°$) (a,e), perpendicular ($\chi = 90°$) (b,f) or tilted ($\chi = 135°$) (c,g) with respect to linearly polarized light, and perpendicular ($\chi = 90°$) (d) to the propagation axis of circularly polarized light of +1 helicity. Each figure is a cut of the MFPAD: in any plane which contains the molecular axis (for $\chi = 0°$), in the half planes $\phi_k = 0°$ and $\phi_k = 180°$ (for both $\chi = 90°$ and $\chi = 135°$), and in the half planes $\phi_k = 90°$ and $\phi_k = 270°$ (for $\chi = 90°$ and circularly polarized light). (a,b,c,d) and (e,f,g) are derived from an experiment performed with elliptically and linearly polarized light, respectively (see text). Experiment and theory are normalized in such a way that the integral PI cross section is identical. The absolute scale (in Mbarn) corresponding to the radius of each figure is indicated near the reference circumference.

VI. RECOIL FRAME PHOTOEMISSION IN PI OF LINEAR MOLECULES: NON-AXIAL RECOIL INDUCED BY ROTATION OR BENDING PRIOR TO DISSOCIATION

In this section we illustrate two different cases of non-axial recoil in DPI of linear molecules. We select as examples two reactions of valence shell ionization of the N_2O molecule that were thoroughly studied.

The valence shell electronic structure of the N_2O molecule in the N_2O(X 1Σ+) linear ground state is (4σ)2(5σ)2(6σ)2(1π)4(7σ)2(2π)4 and the two lowest dissociative ionic states are the N_2O+(B 2Π) and N_2O+(C 2Σ+) electronic states corresponding to ionization of the 1π and 6σ molecular orbitals, respectively. These two PI reactions induced by linearly and circularly polarized light have been studied using the VC method in the 18-22 eV photon excitation energy range, from threshold to a few eV above threshold [72,103]. Several DPI channels have been identified leading to the production of (NO^+,e) and (N_2^+,e) coincident events for PI into the N_2O+(B 2Π) state, plus the two additional (N^+,e) and (O^+,e) channels for PI into the N_2O+(C 2Σ+) state. For the explored energies all these reactions are "complete" in the sense that they dissociate into two fragments, allowing the kinematics of the dissociation to be determined from the momentum of the detected ion fragment; breaking of the N-N bond is more probable than breaking the N-O bond.

The Kinetic Energy Correlation Diagram (KECD) for the (NO^+,e) dominant ionic channel at the $h\nu = 21.2$ eV excitation energy displayed in Fig. **4a** illustrates the population of the $N_2O^+(B\ ^2\Pi)$ and N_2O^+ ($C\ ^2\Sigma^+$) ionic states. Fig. **4b** and **4c** zoom on the characteristics of the PI reaction into the $C\ ^2\Sigma^+$ state analyzed through the detailed electron-ion fragment kinetic energy sharing corresponding to the (NO^+,e) and (N^+,e) ionic channels for $h\nu = 20.5$ eV. The resolved vibrational structure of the N_2O^+ ($C^2\Sigma^+$), consistent with previous PES results, shows that this state is linear with a dominant excitation of the (0,0,0) vibrational state, in contrast with that of the $B\ ^2\Pi$ which is rather complex.

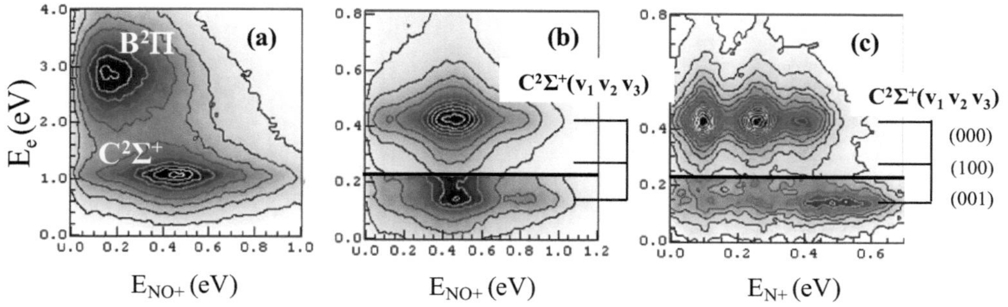

Figure 4: (a) KECD for the (NO^+,e) ionic channel at $h\nu = 21.2$ eV identifying PI into the $N_2O^+(B\ ^2\Pi)$ and $N_2O^+(C\ ^2\Sigma^+)$ ionic states, which reveal different structures. (b) and (c) zoom on the characteristics of PI to the $C\ ^2\Sigma^+$ state analyzed through the KECDs of the (NO^+,e) and (N^+,e) ionic channels at $h\nu = 20.5$ eV. The structures in the (N^+,e) KECD at $E_e \approx 0.4$ eV demonstrate dissociation of the $N_2O^+(C\ ^2\Sigma^+$, 0,0,0) ionic state into the $NO(X^2\Pi;v=0,1,2)$ + $N^+(^3P)$ vibrationally excited channels [103].

VI.A. RFPADs for Valence Shell PI of N_2O into the $N_2O^+(C\ ^2\Sigma^+)$ State

First considering PI into the linear $N_2O^+(C\ ^2\Sigma^+$ (0,0,0)) state the main PI reactions correspond to:

$$\begin{aligned} N_2O\left(X\ ^1\Sigma^+\right)+h\nu &\rightarrow N_2O^+\left(C\ ^2\Sigma^+(0,0,0)\right)+e \\ &\rightarrow NO^+\left(X\ ^1\Sigma^+, v=3\right)+N\left(^2P\right)+e \end{aligned} \quad (49)$$

$$\begin{aligned} N_2O\left(X\ ^1\Sigma^+\right)+h\nu &\rightarrow N_2O^+\left(C\ ^2\Sigma^+(0,0,0)\right)+e \\ &\rightarrow N^+\left(^3P\right)+NO\left(X\ ^2\Pi, v=0,1,2\right)+e \end{aligned} \quad (50)$$

Taking into account that the molecular axis orientation is reversed when considering equation (49) and (50) a very similar $I(\theta_k, \phi_k, \chi)$ angular distribution is obtained for both reactions. The five $F_{N,0}^{(L)}$ functions identical to the $F_{LN}(\theta_k)$ functions [9,11], extracted from the Fourier analysis of the $I(\theta_k, \phi_k, \chi)$ distribution according to the general expression given in section III, are displayed in Fig. **5a**. For the purpose of the present discussion, 3D plots of the recoil frame photoemission patterns are only displayed for the two molecular orientations of the NO^+ recoil axis parallel, $I_{\chi=0}(\theta_k)$, and perpendicular, $I_{\chi=90}(\theta_k, \phi_k)$, to the polarization axis of linearly polarized light which feature the parallel and perpendicular components of the PI transition, and the polar angle θ_k is defined relative to the O end of the N_2O molecular axis.

We note that throughout this chapter, for clarity, the 3D plots $I_\chi(\theta_k, \phi_k)$ describing the measured photoemission patterns do not indicate the uncertainty attached to the measurement: however, the evaluation of error bars are given in the report of the $F_{N,0}^{(L)}$ functions and that of the 1D cuts of the MFPADs or RFPADs, as shown *e.g.* in Fig. **5a**. The measured $F_{LN}(\theta_k)$ functions display smooth oscillations with a limited contrast as compared with previously studied reactions involving well defined initial and final

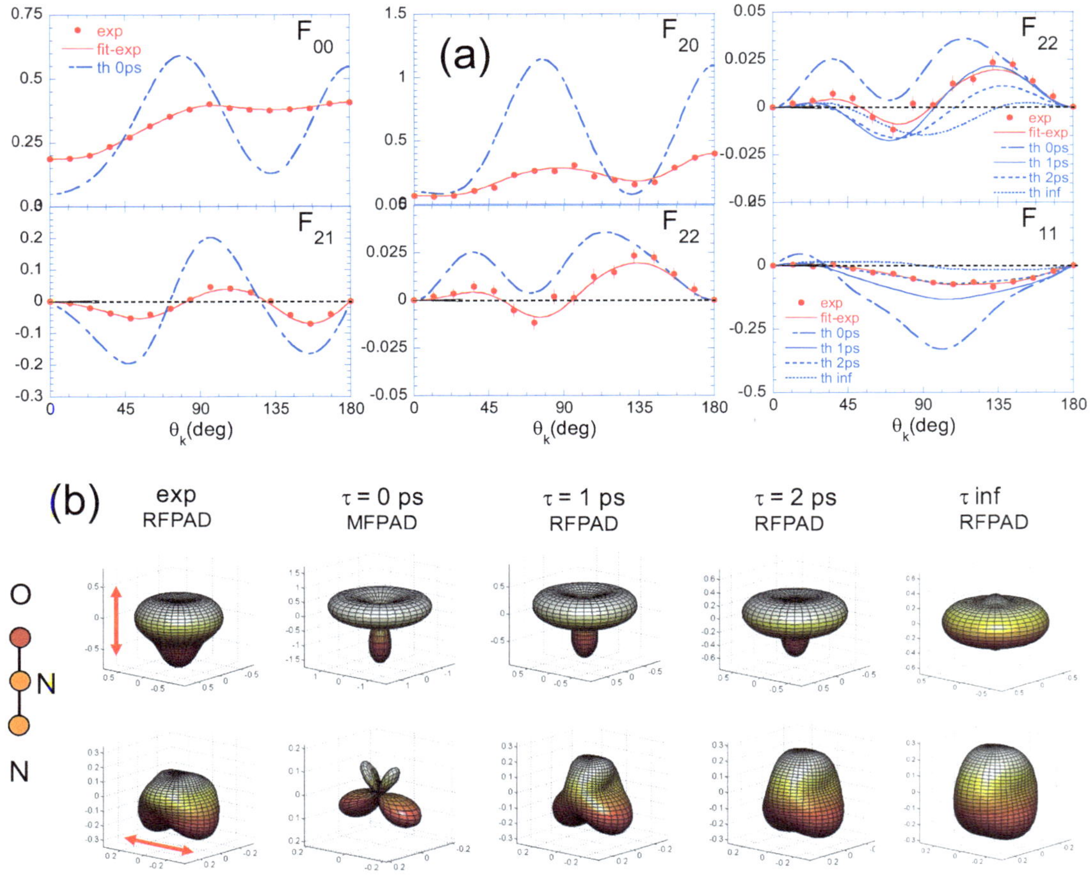

Figure 5: (a) Measured (dots) F_{LN} functions for DPI into the $N_2O^+(C\ ^2\Sigma^+)$ state at $h\nu$ = 20.9 eV compared with MCSCI calculations (dashed line) assuming a prompt dissociation (τ = 0 ps). On the right column F_{22} and F_{11} are computed for different lifetimes of the C states as shown. (b) Measured RFPADs for the parallel and perpendicular orientations of the molecular axis with respect to linearly polarized light compared with the computed MFPADs (τ = 0 ps) and RFPADs for τ ≠ 0. A fair agreement is obtained for a predissociation time of the $N_2O^+(C\ ^2\Sigma^+)$ state for values near 2 ps.

electronic states [9,11]. Accordingly, although the related measured RFPADs display significant emission anisotropies, different for the two orientations, notably a preferred electron emission along the molecular axis in the direction of the N end of the molecule for the parallel transition, their shape is rather smooth. It is also noticed that the F_{22} function, which is expected to take only positive values since the PI reaction involves a Σ^+ (initial neutral state) $\rightarrow\Sigma^+$ (final ionic state) transition [9], becomes negative for some electron emission directions θ_k. The linearity of the $N_2O^+(C\ ^2\Sigma^+$ (0,0,0)) state, combined with the characteristics of recoil frame photoemission, can only be understood if one assumes a finite predissociation time of the $C\ ^2\Sigma^+$ state, prior to dissociation breaking the N-N bond. The measured F_{LN} functions and RFPADs are compared with MCSCI calculations in Fig. **5a**, assuming different lifetimes in the picosecond regime (see equation (26)).

The five channel MCSCI calculation is described in ref. [103]. For τ = 0 ps, the computed F_{LN} functions display oscillations in phase with the measured values, however the contrast is significantly larger: the MFPADs for the parallel and perpendicular transitions are highly structured and show the dominant role of $d\sigma$ and $d\pi$ partial waves in the scattered electronic wave function. The strong ϕ_k azimuthal anisotropy, where emission takes place in the plane defined by the molecular and polarization axes, is characteristic of the $\Sigma^+ \rightarrow \Sigma^+$ character of the PI transition. In the computed F_{LN} functions and RFPADs including rotation of the molecular axis prior to dissociation according to equation (26), the rotational temperature of the N_2O is fixed to 35K, a reasonable estimate for the type of supersonic expansion used. A fair agreement between the measured and computed F_{LN} functions and RFPADs is obtained for lifetimes near 2 ps. In this example, the detailed combined experimental and theoretical study provides both information about the PI dynamics for ionization into the $N_2O^+(C\ ^2\Sigma^+)$ state and the nuclear dynamics of the dissociation reaction, in the simplest case of rotation of a linear molecule. Similar examples have been studied for valence and inner-valence ionization involving excited electronic states of diatomic molecules such as O_2 [10] and CO.

VI.B. RFPADs for Valence Shell PI of N_2O into the $N_2O^+(B\ ^2\Pi)$ State

Ionization into the $N_2O+(B\ ^2\Pi)$ state features another type of combined PI and dissociation dynamics leading to rather smoothed F_{LN} functions and RFPADs. Investigation of this ionization process by two-photon absorption spectroscopy led to an estimated lifetime of the $B\ ^2\Pi$ ionic state of 3×10^{-14}s. This value, much shorter than the rotational period of the N_2O molecule, rules out rotation as a source of a breakdown of the axial recoil approximation. On the other hand, the PES has shown that ionization into the $B\ ^2\Pi$ state gives rise to a complicated vibrational pattern extending over 1.5 eV, consistent with the KECD displayed in Fig. **4a**. This complex structure could be tentatively assigned to vibronic coupling of two electronic states of Π symmetry, as well as to the role of a bent character of the $B\ ^2\Pi$ state, the latter assumption being explored in this work. The PI dynamics a few eV above threshold ionization to the $B\ ^2\Pi$ state is also strongly influenced by resonant excitation of Rydberg series converging to the $N_2O+(C\ ^2\Sigma+)$ state. In order to simplify the study of the conditions of axial or non-axial recoil for this DPI process, we select photon excitation energies between resonant states for which PI into the $N_2O+(B\ ^2\Pi)$ state is governed by direct ionization as reported in [72]. The PI dynamics attached to interferences between resonant and non-resonant ionization has been discussed specifically through the study of the circular dichroism observable [15]. The dominant DPI reaction is assigned to:

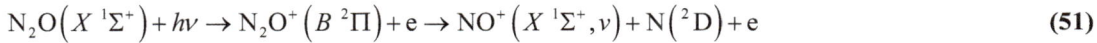

$$N_2O\left(X\ ^1\Sigma^+\right) + h\nu \rightarrow N_2O^+\left(B\ ^2\Pi\right) + e \rightarrow NO^+\left(X\ ^1\Sigma^+, v\right) + N\left(^2D\right) + e \qquad \textbf{(51)}$$

where the vibrational distribution of the $NO^+(X\ ^1\Sigma^+, v)$ state extends from $v = 0$ to $v = 6$, with a maximum corresponding to $v = 4$. Fig. **6** displays the measured RFPADs, here again limited for simplicity to the two orientations of the molecular axis parallel and perpendicular to the polarization axis of linearly polarized light.

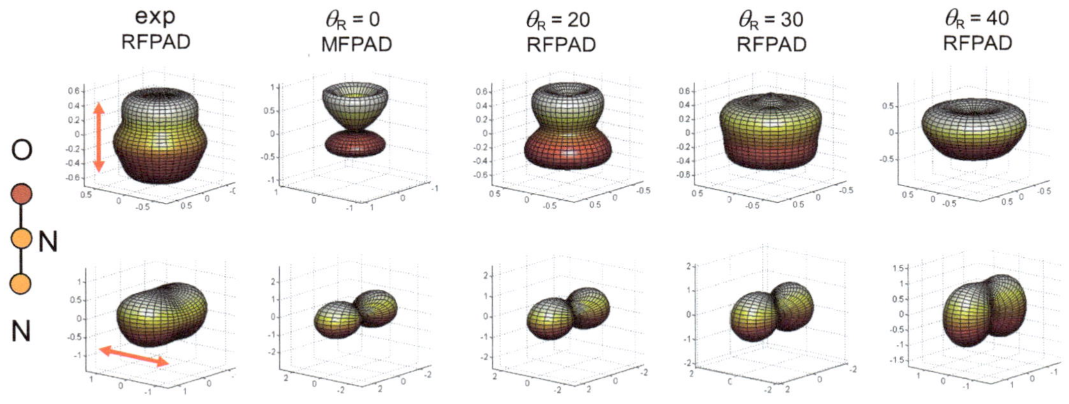

Figrue 6: Measured RFPADs for DPI into the $N_2O^+(B\ ^2\Pi)$ state at $h\nu = 18.65$ eV for the parallel and perpendicular orientations of the molecular axis with respect to linearly polarized light, compared with the computed MFPADs (θ_R = 0°) and RFPADs for different recoil angles $\theta_R \neq 0$. A fair agreement is obtained for a recoil angle of the NO^+ fragment of about 30° *i.e.* a bent geometry of the $N_2O^+(B\ ^2\Pi)$ state which corresponds to a $\angle$ O-N-N angle of about 120°.

The main remarkable feature is the preferred electron emission in a direction perpendicular to the plane containing the polarization and molecular axes for the perpendicular transition, for the low photoelectron energies, a fingerprint of a $\Sigma^+ \rightarrow \Pi$ PI reaction, as observed *e.g.* in PI of O_2 [10] and discussed in detail in [72]. The measured RFPADs are compared to the MFPADs and RFPADs computed using the MCSCI calculation as performed for PI into the $C\ ^2\Sigma^+$ state, also displayed in Fig. (**6**), and accounting for bending of the $N_2O^+(B\ ^2\Pi)$ state after ionization and prior to dissociation as formalized in section III equations (17 and 29). The computed RFPADs correspond to different values of the recoil direction labeled by the angle θ_R defined with respect to the initial molecular axis. A reasonable description of the experimental findings is obtained for a recoil direction of about $\theta_R \approx 30°$, which corresponds to a bending angle of about $\theta_B \approx 120°$ for the quasi-symmetric N_2O molecule. Further extensions in the calculation could be explored: the *B* state can be treated as a vibronic state that includes more than one electronic state rather than as a single electronic state; a second extension of the calculation would be explicitly examine the dynamics of dissociation to extract the recoil angle or angular distributions. Finally, the expansion of the continuum state could be extended to include more ion states in particular at the relatively low photoelectron energies considered.

VII. MFPADs and RFPADS in L SHELL IONIZATION OF THE C_{3v} CH_3Cl molecule

As stated above, in non-linear polyatomic molecules, when only two fragments are produced in dissociative ionization, it is not possible to measure the MFPAD and only RFPADs are accessible, even when the axial recoil approximation is valid. The experimental data are then averaged over unobserved molecular angles [4,7]. Here we first discuss the fingerprints of the computed MFPADs, then compare the computed and measured RFPADs.

Examples of RFPADs from a non-linear molecule are illustrated in the study of the ionization of the Cl 2*p* orbital in molecules such as the CH_3Cl methylchloride [70,94]. This reaction also features specific properties of molecular frame photoemission where the photoelectron is ionized from quasi-degenerated molecular orbitals of different symmetry. In the photoelecton spectrum one finds two ion states that correspond to the $^2p_{1/2}$ and $^2p_{3/2}$ atomic states of the Cl atom, with Ionization Potential (IP) of 207.9 and 206.3 eV, respectively [104]. In the full molecular symmetry there would actually be three states with symmetry $E_{1/2}$, $E_{1/2}$, and $E_{3/2}$ [105], however in this system the splitting due to the non-spherical environment (0.05 eV from Koopmans's IPs) is much smaller than the spin-orbit splitting of 2.6 eV. Thus the final states will be very close to the atomic states and one of the $E_{1/2}$ states will be nearly degenerate with the $E_{3/2}$ state. In both the $^2p_{1/2}$ and $^2p_{3/2}$ states the hole has a 1:2 ratio of probabilities for being in the A_1 and *E* states so that we have computed the photoionization for the A_1 and *E* states separately and just added the resulting MFPADs and RFPADs together to compare with the measured angular distributions for the Cl $^2p_{1/2}$ and $^2p_{3/2}$ states. Fig. **7** displays the MFPADs for three selected orientations of the polarization, parallel to the C-Cl molecular axis *z*, and perpendicular to the C-Cl molecular axis considering the two different cases where the polarization is in the *xz* plane which contains a CH bond with the H atom at a positive value of *x*, or perpendicular to this plane, computed using the SC-FCHF type calculation for a photon energy of 211 eV.

The results for the $(2p)^{-1}$ A_1 state and the $(2p)^{-1}$ *E* state are given separately in Fig. **7**. The sum of the two which then corresponds to the MFPAD for either the Cl $^2p_{1/2}$ or $^2p_{3/2}$ states is displayed in Fig. **8**.

In the general form of the MFPAD for linearly polarized light given in equation (23) involving in principle an unlimited number of $F_{N,\nu}^{(L)}$ functions, the $C_{3\nu}$ symmetry imposes the constraint that all terms except those where ν is a multiple of three are zero, leading to the reduced form:

$$T_{i,f}^{(\mathrm{LP})}(\theta_k,\phi_k,\chi_{\mathrm{LP}},\gamma_{\mathrm{LP}}) = \sum_{L=0,2}\sum_{N=0}^{L}\sum_{k=0,\pm1,\pm2..} F_{N,3k}^{(L)}(\theta_k) \times P_L^N(\cos\chi_{\mathrm{LP}})\cos\left[(N+3k)\phi_k - N\gamma_{\mathrm{LP}}\right] \quad (52)$$

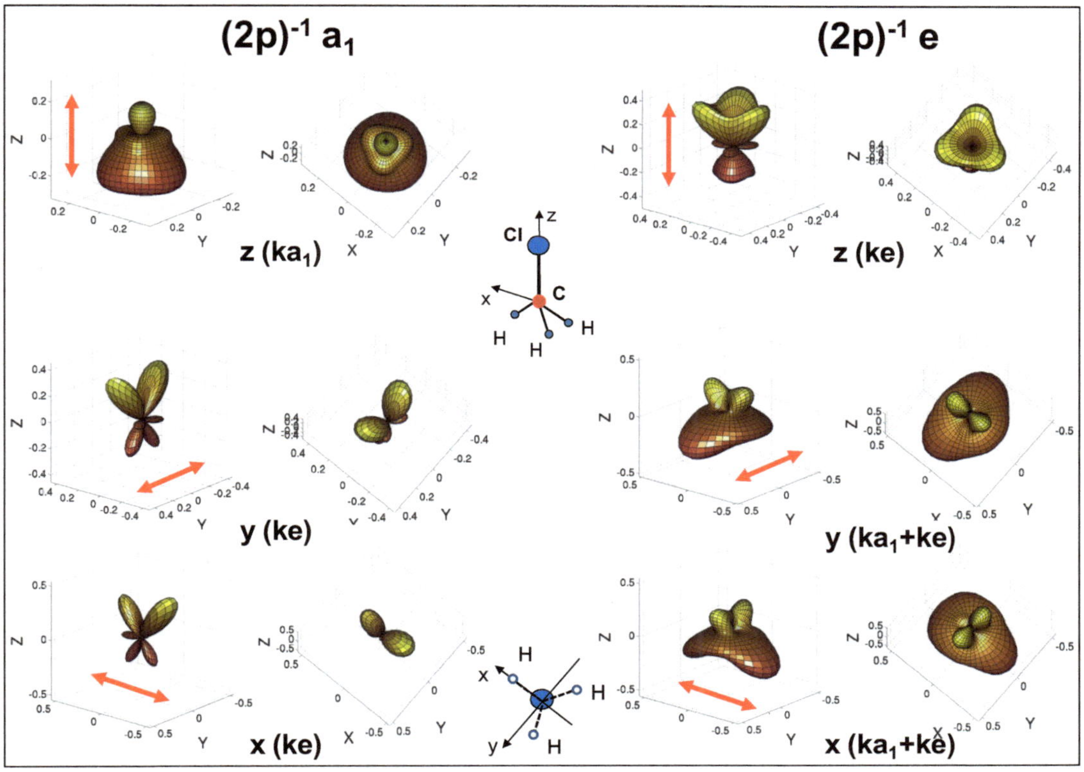

Figure 7: Side and top views of the MFPADs for the ionization of CH_3Cl Cl(2p)$^{-1}$ for both a$_1$ and e symmetries by linearly polarized light at a photon energy of 211eV and a photoelectron energy of 5 eV. The *xyz* molecular frame is chosen such that *z* is along the CCl axis and one CH bond lies in the *xz* plane, with the H atom at a positive x. The polarization axis is parallel, or perpendicular to the CCl axis along *x* or *y*. The symmetry of the electronic wavefunction in the ionization continuum is indicated in parentheses for each transition moment.

The three-fold symmetry of the CH_3Cl molecule is most clearly seen for the parallel excitation ($\chi = 0°$). The three lobes on top of the distribution for the (2p)$^{-1}$ a$_1$ ionization point towards the directions that the H atoms are pointed, whereas for the (2p)$^{-1}$ e ionization they point between these directions.

If this were just the Cl atom, one would have the same PAD in the positive *z* direction as in the negative *z* for both the parallel polarization case shown in the top row and the perpendicular ionization shown in the bottom rows of Fig. **7**. In the parallel ionization of the 2*p* A_1 orbital there is a strong asymmetry indicating the importance of scattering in the final state in determining the form of the MFPAD. For the 2*p* *E* orbitals, there is less asymmetry for parallel ionization and it is reversed. This reduction in the importance of final state scattering effects with the 2*p* *E* orbitals can be attributed to the fact that the C atom is in the nodal plane of the outgoing *E* symmetry scattered waves for ionization by light aligned along the C-Cl axis. The perpendicular excitation in the (2p)$^{-1}$ e ionization displays noticeable features and scattering asymmetries which may also be assigned to such scattering effects. The MFPAD is composed of a dominant oblong contribution rather well aligned along the polarization and a smaller two-lobe structure which is contrastingly oriented at 90° with respect to the polarization. Consider *e.g.* *x* polarization, where the primary intensity comes from ionization from the p_x orbital: it leads to an angular dependence of the form x^2 which in the C_{3v} symmetry can be decomposed into an a_1 component, x^2+y^2 and an *e* component x^2-y^2. When these two terms have the same phase, one recovers the original x^2 angular dependence as reflected by the dominant part of the MFPAD; however the smaller part in the positive *z* direction comes from scattering from the CH_3 group, which will lead to different phase in the x^2+y^2 and x^2-y^2 components. The observed orientation is consistent with a π phase shift resulting in an y^2 form after superposition of the two waves.

In the MFPAD obtained from the sum of the two symmetries, presented in the left column of Fig. **8**, the largest resulting asymmetry is in the ionization by light with the perpendicular orientation. A detailed analysis of the partial wave composition of the dipole matrix elements shows that the perpendicular transition for the $2p$ E orbitals is dominated by $l = 2$ contributions.

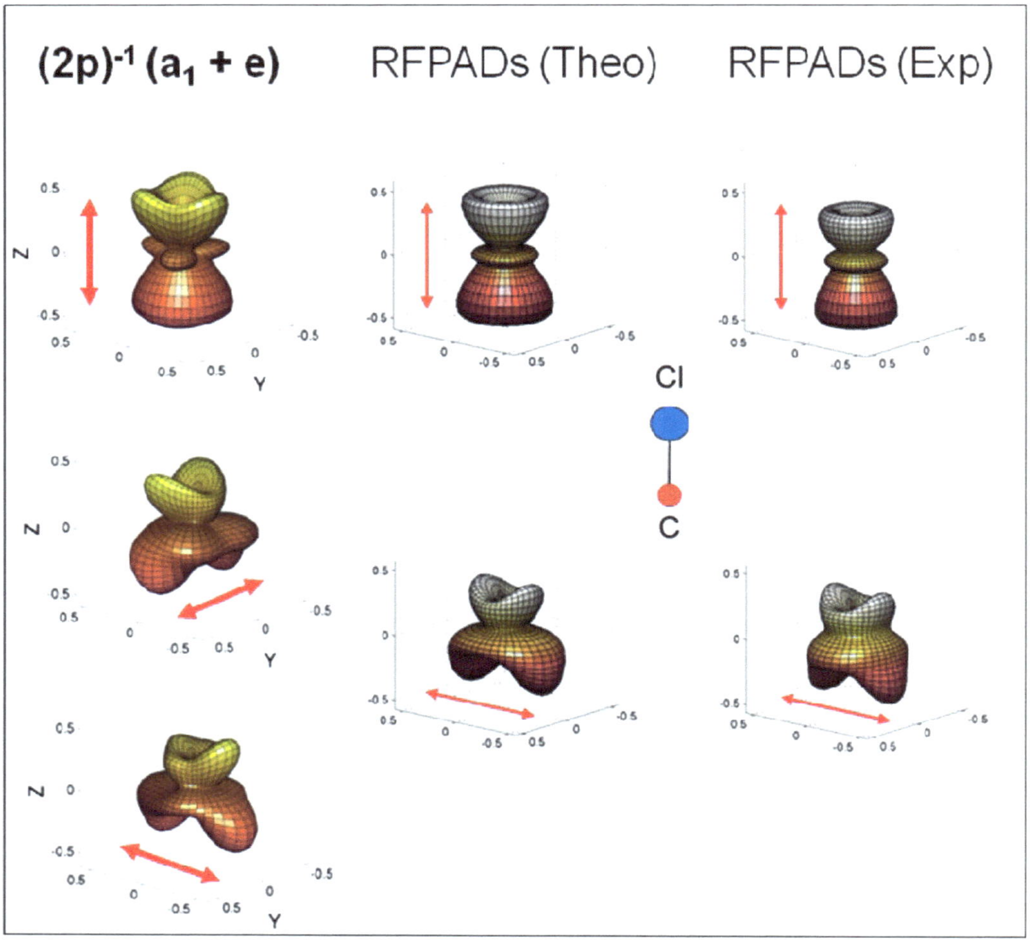

Figure 8: Sum of the MFPADs corresponding to the $(2p)^{-1}$ A_1 state and the $(2p)^{-1}$ E states displayed in Fig. (**7**) (left column) for the three orientations of the polarization. Computed (middle) and measured (right column) RFPADs for a polarization direction parallel and perpendicular to the recoil axis V_{Cl+} (top).

We now consider the dominant channel measured using the VC method and elliptically polarized light at $h\nu$ = 211 eV at BESSY, where detection of the CH_3^+ and Cl^+ fragments in coincidence with the initial photoelectron is the signature of the breaking of the C-Cl bond along the C_{3v} z axis, subsequent to Auger decay. This channel features the simplest case of RFPADs since it averages only over the unobserved azimuthal angle which gives the orientation of the molecule about the recoil axis, washing out the information about the locations of the H atoms.

We focus the discussion of the RFPADs obtained for the two orientations of the recoil axis parallel and perpendicular to the axis of linearly polarized light. The details of the reaction involving intermediate states of the dication, as well as the circular dichroism effects will be discussed elsewhere.

In Fig.**8** we present a comparison of the measured RFPADS and those computed assuming the validity of the axial recoil approximation. The computed RFPAD is obtained from the MFPAD shown in the left column by merely averaging over the orientation of the H atoms as implied by equation (35). The three-dimensional plots shown in Fig. **8**, demonstrate a remarkable agreement between experiment and theory for this system. This indicates that for this core ionization process, the SCFCHF approximation used here can give a very good representation of the process, meaning that electronic correlation plays a negligible role.

Despite the suppression of the structures characterizing the presence of the H atoms, we observe that the dominant features of the MFPADs can still be identified in the RFPADs. They can be well assigned to the fingerprints of the MFPADs for ionization of both the a_1 and e orbitals discussed in Fig.**7**. The situation is quite different when a fragmentation channel such as (H^+, CH_2Cl^+) is considered: this pertains to the case where the recoil axis is not a symmetry axis of the molecule, illustrated in section VIII by the example of PI of NO_2.

The VC method gives access to more complex fragmentation channels through the record of multiple coincident events such as those including the photoelectron and the three ion fragments (H^+, CH_2^+, Cl^+) produced after double Auger decay. This channel is well characterized and it provides a closer access to the MFPAD since the *xz* plane of the molecular frame can be determined by detection of the H^+ fragment. A similar level of information was reported recently for 3D mapping of photoemission induced by O 1s ionization of the C_{2v} H_2O bent molecule, followed by double Auger decay, analyzing the photoelectron in coincidence with the three ions (O^+, H^+, H^+) [33] leading in that case to the complete MFPADs. The KER of the (H^+, CH_2^+, Cl^+) ions is about 17 eV and the measured emission angles of the fragments $\Theta_{Cl+\text{-}H+} \approx 113°$, $\Theta_{Cl+\text{-}CH2+} \approx 160°$, $\Theta_{H+\text{-}CH2+} \approx 87°$ are rather comparable to the bond directions in the molecule, corresponding to a break-up of the trication under close-to-axial-recoil conditions. The expansion equation (52), limited to $k = 0$ and $k = \pm 1$, involves ten $F_{N,3k}^{(L)}(\theta_k)$ functions, which can be extracted from a fit of the measured $T_{i,f}^{(\text{LP})}(\theta_k, \phi_k, \chi_{\text{LP}}, \gamma_{\text{LP}})$ distribution, providing the basis for determining the MFPADs. For example, the MFPAD for the parallel transition, with $\chi_{\text{LP}}= 0°$ and $\gamma_{\text{LP}}= 0°$, is simply obtained as:

$$T_{i,f}^{(\text{LP})}(\theta_k, \phi_k) = \left[F_{0,0}^{(0)}(\theta_k) + F_{0,0}^{(2)}(\theta_k)\right] + \left[F_{0,3}^{(0)}(\theta_k) + F_{0,3}^{(2)}(\theta_k)\right]\cos(3\phi_k) \quad \textbf{(53)}$$

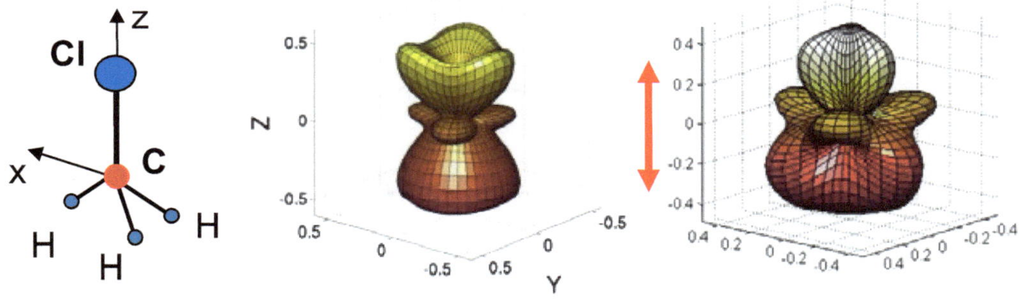

Figure 9: Measured (right panel) MFPAD for an orientation of the polarization axis parallel to the C-Cl bond derived from the analysis of the (H^+, CH_2^+, Cl^+, e_{ph}) events compared with the computed (middle panel) MFPAD.

Fig. **9** illustrates the measured MFPAD for the parallel transition, which compares reasonably well with the computed one presented in Fig. **8**. Although the statistics for this channel was rather low, the data analysis method relying on the general expression given in equation (53) enabled us to extract the main features of the MFPAD.

VIII. MFPADs and RFPADs in valence SHELL IONIZATION OF THE C_{2v} NO_2 molecule

We will next consider a case where the recoil axis is not a symmetry axis of the system. In such a situation, the RFPAD can be expected to be much less structured due to the loss of symmetry. To look at these effects, we will consider the $(1a_2)^{-1}$ process in PI of NO_2 [73]. NO_2 is a bent molecule with C_{2v} symmetry. The MFPAD computed in the SCFCHF approximation for ionization from the $1a_2$ orbital at a photon energy of 15.7 eV is displayed in Fig. **10**, for light linearly polarized in the *x*, *y*, and *z* directions leading to electrons in the kb_2, kb_1, and ka_2, continua, respectively.

In this figure, the molecule is in the *yz* plane and the N atom below the O atoms in the *z* direction. The shape of the MFPADs can be qualitatively understood by a simple consideration of the angular momentum composition of the initial orbital and the angular momentum and corresponding nodes contributed by the various light polarizations [34]. The $1a_2$ initial orbital can be best described as an xy orbital formed from the lone-pair $2p$ orbitals on the O atoms that are perpendicular to the plane containing the molecule. For the x-polarized light, the outgoing wave should be predominately x^2y in character and that is reflected in the strong peaks in the intensity in the $\pm\hat{y}$ directions. The shape for the z polarized light is clearly seen as an $l = 3$ distribution coming from the xyz expected symmetry obtained from the product of the orbital and photon angular momenta. The one surprising MFPAD is that found in the case of the y polarized light. In this case the expected xy^2 symmetry is not seen, instead a strong xz symmetry is found. This must then be attributed to strong scattering of the photoelectron by the molecular ion potential.

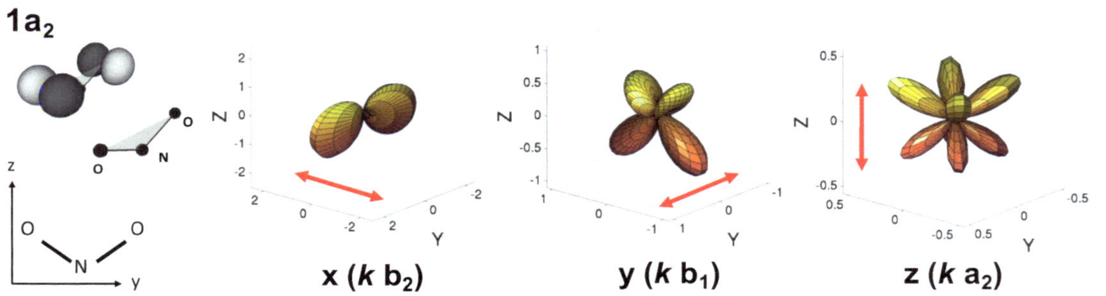

Figure 10: MFPADs computed for PI of the $1a_2$ molecular orbital of NO_2 induced by linearly polarized light parallel to the x, y, and z axes of the C_{2v} symmetry at $h\nu$ = 15.5 eV. z is the main symmetry axis and yz corresponds to the molecular plane.

The RFPADs were obtained experimentally using the VC method with circularly polarized light, selecting direct ionization into the $NO_2^+(b\,^3A_2)$ ionic state according to the reaction:

$$\begin{aligned} NO_2\left(X\,^3A_2\right) + h\nu &\rightarrow NO_2^+\left(b\,^3A_2\right) + e \\ &\rightarrow NO^+\left(X\,^1\Sigma^+, v=0\right) + O\left(^3P\right) + e \end{aligned} \quad (54)$$

A comparison of the computed RFPADs and the experimentally observed angular distributions requires the choice of a recoil direction relative to the molecular axes and an averaging over the azimuthal angle about the recoil direction leading to the functional form of the RFPAD given in equations (34) and (35). If the fragmentation were strictly to follow the axial recoil approximation, assuming the ground state bond angle of 134°, one would expect a recoil direction (α_R, β_R) relative to the symmetry axis of the initial C_{2v} symmetry of the NO_2 molecule of ($\alpha_R = 90°$, $\beta_R = 113°$) for an NO^+ fragment, and ($\alpha_R = 90°$, $\beta_R = 67°$) for an O^+ fragment coming from direct dissociation of an NO bond in the molecular frame.

In Fig. **11** we report the measured values of the $F_{N,0}^{(L)}$ functions and the computed values obtained with α_R = 90° and β_R = 110°, 120°, and 130° for ionization leading to the $(1a_2)^{-1}\ b^3A_2$ state of NO_2^+ at $h\nu$ = 15.7 eV, corresponding to a photoelectron of 2.1 eV. We find fairly good agreement between the computed and measured $F_{N,0}^{(L)}$ functions. The values of the $F_{0,0}^{(0)}$ and $F_{0,0}^{(2)}$ functions are most sensitive to the recoil angle at θ_k = 0° and 180°. At those angles, we can see that the value of $\beta_R \approx 120°$ provides the best overall fit to the experiments. The general behavior and relative magnitude of the measured $F_{1,0}^{(2)}$ and $F_{2,0}^{(2)}$ functions are very well predicted by the calculations. The negative values of the $F_{0,0}^{(2)}$ function reflect the dominant "perpendicular character" of the $1a_2 \rightarrow kb_2$ transition for ionization into the NO_2^+ ($b\,^3A_2$) state, corresponding to a polarization axis perpendicular to the molecular plane, and therefore perpendicular to the ion fragment recoil axis. The negative sign of $F_{2,0}^{(2)}$ indicates a preferred electron emission perpendicular to the plane defined by the recoil ion direction and the polarization axis (ϕ_k = 90° and 270°).

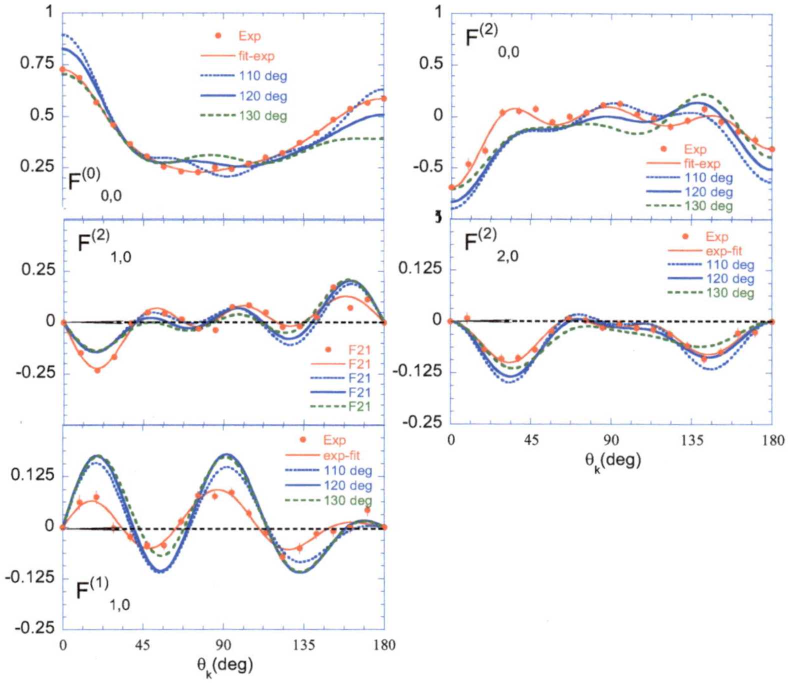

Figure 11: Measured (dots) $F_{N,0}^{(L)}$ functions for one-photon PI of NO_2 induced by circularly polarized light at $h\nu = 15.7$ eV, compared with computed values corresponding to a recoil NO^+ direction $\alpha_R = 90°$ and $\beta_R = 110°$, 120°, and 130° (dashed and full line as shown).

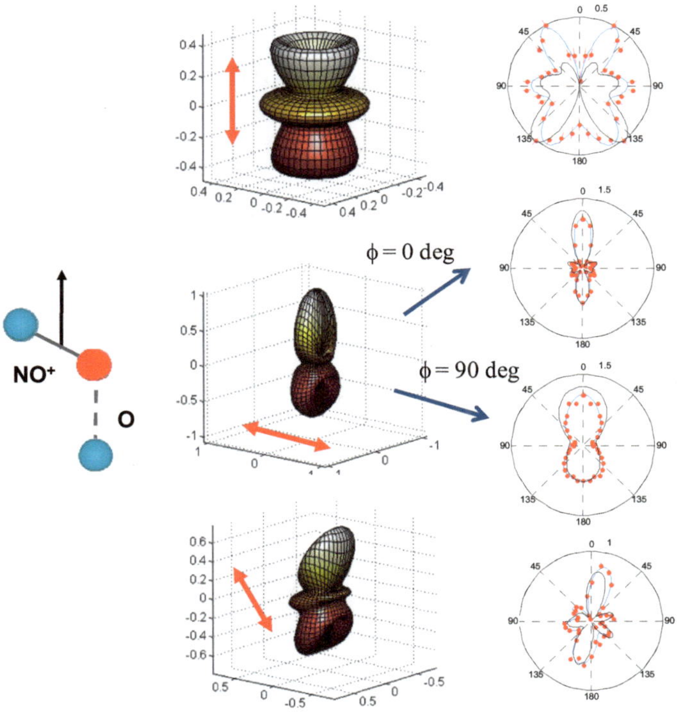

Figure 12: Measured RFPADs at $h\nu = 15.7$ eV where the recoil axis is oriented parallel, perpendicular, and at 45° with respect to the axis of linearly polarized light. The relevant cuts of the RFPADS are shown on the right column and compared with SCFCHF calculations.

Fig. **12** displays the measured and computed RFPADs at $h\nu = 15.7$ eV where the recoil axis direction in the laboratory frame is oriented parallel, perpendicular, and at 45° with respect to the light polarization axis. The shapes of the RFPADs reveal significant anisotropies, consistent with the reported $F_{N,0}^{(L)}$ functions: the most probable transition corresponds to the case where the ion recoil axis is perpendicular to the polarization, and electron emission takes place dominantly in the molecular plane. It reflects the dominant $a_2 \rightarrow kb_2$ transition and x^2y character of the MFPAD as shown in Fig. **10** for x-polarized light, with strong electron emission intensities in the $\pm\hat{y}$ directions. Experimental and theoretical results are in very good agreement, except for the emission anisotropy for the parallel orientation of the recoil axis which is found to be larger in the calculation. We note that in such a case photoemission asymmetries are induced by the ion fragment recoil direction tilted with respect to the symmetry axes of the molecule.

IX. MULTI-PHOTON FRAGMENTATION AND IONIZATION OF THE NO_2 MOLECULE

We finally address in this section, as an example to illustrate the methodology extended to multi-photon ionization processes, the study of multi-photon excitation of the NO_2 molecule induced by the absorption of four or five photons of 400 nm wavelength, delivered by SLIC linearly and circularly polarized femtosecond laser sources described in section II.A, using the VC method. Although the analysis of the PADs would be simpler for a linear molecule, we first studied NO_2 since there is such an extensive literature on the challenging photophysics of this system. The photodynamics in NO_2 is complicated by numerous non-adiabatic couplings and has motivated a number of multiphoton and time-resolved studies [46,49,106]. In the reported experiments, three main channels producing ionic species are observed and assigned to (*i*) four-photon ionization into the $NO_2^+(X\ ^1\Sigma_g^+)$ ground state, (*ii*) five-photon dissociative photoionization (DPI) into $(NO^+(X^1\Sigma^+,v) + O(^3P) + e)$ and (*iii*) four-photon induced $NO^+(X^1\Sigma^+,v) + O^-(^2P)$ ion-pair formation. The characteristics of the ionization reactions, photoelectron energy spectra for non-dissociative ionization and (E_e, E_{NO+}) KECDs for dissociative photoionization, strongly differ from those observed for one-photon ionization at comparable photon excitation energies [73]. Compared to one-photon ionization, the multiphoton excitation accesses different Franck-Condon factors *via* resonant intermediate states and changes the probability of multielectron excitation processes. In the multi-photon excitation scheme, the ion-pair formation is enhanced by one order of magnitude [107]. DPI and ion-pair formation reveal similar features such as a significant vibrational excitation of the $NO^+(X^1\Sigma^+,\ v = 0\text{-}5)$ molecular fragment and a recoil ion fragment emission strongly aligned along the polarization axis of linearly polarized light, or preferentially emitted in the plane perpendicular to the propagation axis of circularly polarized light.

Ion-pair formation corresponds to a bound-to-bound 4-photon channel populating a super-excited state of the NO_2 molecule, above the $NO_2^+(X\ ^1\Sigma_g^+)$ adiabatic threshold: the formalism describing the recoil anisotropy for bound-to-bound n-photon transition inducing prompt axial recoil dissociation of a non-linear molecule presented in section III.C has been developed to interpret the measured anisotropies in terms of excitation pathways *via* near-resonant intermediate states of given symmetries. We briefly describe here the measured recoil fragment angular distribution induced by linearly polarized light, compared with bound-to-bound expressions of the excitation probability, i. e. $H_v^{(n,\lambda,\delta)}(\chi,\beta)$ in equation (39) associated with different reaction pathways. For linearly polarized light and the NO_2 ground state geometry, $H_v^{(n,\lambda,\delta)}(\chi,\beta)$ reduces to the $H_v^{(n=4,\mathrm{LP})}(\chi_{\mathrm{LP}})$ function. Selected examples relevant for the reaction studied here are

$$H_{fi}^{(4,\mathrm{LP})}[2,2,2,4] = -0.1077\cos^8(\chi) + 0.6040\cos^6(\chi) + 0.1089\cos^4(\chi) + 0.0011\cos^2(\chi) \tag{55}$$

$$H_{fi}^{(4,\mathrm{LP})}[2,2,2,2] = -0.3737\cos^8(\chi) + 0.4964\cos^6(\chi) + 0.3641\cos^4(\chi) + 0.0262\cos^2(\chi) + 0.00015 \tag{56}$$

where the digits in brackets describe a type-1 (x), type-2 (y) or type-3 (z) transition between states of C_{2v} symmetry corresponding to *e.g.* $A_1 \rightarrow B_1$, $A_1 \rightarrow B_2$ and $B_2 \rightarrow B_2$ transitions, respectively. A type-4 transition implies a transition moment that is directed along the dissociating bond. This might occur either because the molecule has become linear or because the two NO bond lengths have become different allowing transition moments to be in directions that differ from those allowed in C_{2v} structures. In the excitation of NO_2 we also examined three-transition pathways involving a $B_2 \rightarrow B_2$ two-photon (2p) transition, labeled as 2p-5 [107], computing two-photon transition moments using MOLPRO [108] and a five-state SA-CASSCF calculation at the geometry ∠ O-N-O = 120°. Combining this computed transition moment with *e.g.* two type-2 transitions leads to the additional [2,2,2p-5] four-photon H_{fi} function:

$$H_{fi}^{(4,\mathrm{LP})}[2,2,2\text{p-}5] = -0.7491\cos^8(\chi) + 0.9751\cos^6(\chi) + 0.1363\cos^4(\chi) - 0.0147\cos^2(\chi) + 0.00034 \quad \textbf{(57)}$$

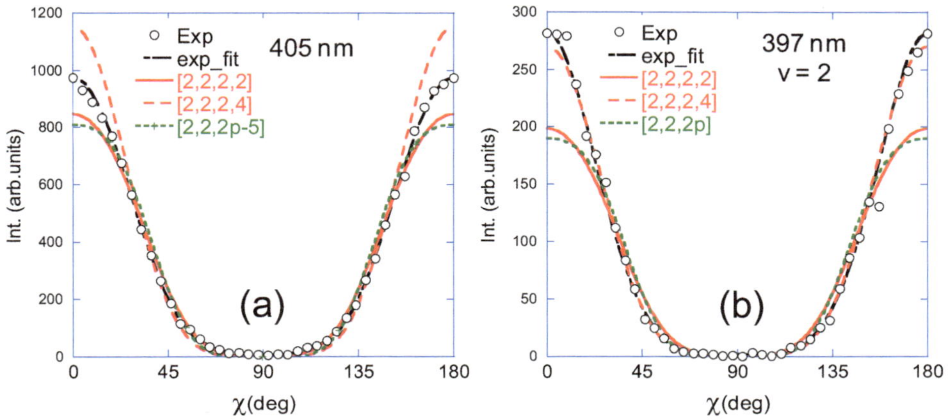

Figure 13: $H(\chi)$ recoil fragment distributions at 405 nm integrated over all KERs (a) and 397 nm for selection of the $NO^+(X^1\Sigma^+, v=2) + O^-(^2P)$ channel (b) for linearly polarized light: exp. (dots and fit black line), model angular distributions for selected four-photon excitation pathways as indicated in the legend (full and dashed lines). The measured and model angular profiles in arbitrary units are normalized such that their total cross sections are identical.

The $H_\nu^{(n=4,\mathrm{LP})}(\chi_{\mathrm{LP}})$ angular profile models involving one type-1 or type-3 transition are clearly excluded by comparison with the measured recoil fragment angular distributions reported in Fig. **13**. The measured anisotropies compared with different models involving type-2 and type-4 transitions shown in Fig. **13** illustrate the sensitivity to the reaction pathway.

We have then explored possible intermediate states that could contribute to the five photon DPI process for which we have obtained experimental RFPADs. In this discussion, we emphasize and disentangle the different elements which contribute to the characteristics of the final RFPAD.

For linear polarization and non-chiral molecules equation (40) gives for the multiphoton RFPAD

$$T_{fi}^{(n,\mathrm{LPRF})} = \sum_{\nu=-2(n-1)}^{2(n-1)} H_\nu^{(n-1,\mathrm{LP})}(\chi_{\mathrm{LP}}) \sum_{L=0,2} \sum_{N=0}^{L} F_{N,\nu}^{(L)}(\theta_k) \times P_L^N(\cos\chi_{\mathrm{LP}}) \cos[(\nu+N)(\phi_k - \gamma_{\mathrm{LP}})] \quad \textbf{(58)}$$

where the $H_\nu^{(n-1,\lambda,\delta)}(\chi,\beta)$ are components of the bound-to-bound part of the excitation probability $T_{i,f}^{(n-1)}(\gamma,\chi,\beta)$ as defined in equation (39). If these functions are known, then it will be possible to extract information about the $F_{N,\nu}^{(L)}$ for non-zero values of ν from experimentally measured recoil-frame multi-photon dissociative photoelectron angular distributions. In the case of five photon DPI of NO_2 considered here, the sum over ν is limited by the number of photons to satisfy $-8 \le \nu \le 8$ so that equation (58) becomes

$$T_{fi}^{(5,\mathrm{LPRF})} = \sum_{\nu=-8}^{8} H_{\nu}^{(4,\mathrm{LP})}(\chi) \sum_{L=0,2} \sum_{N=0}^{L} F_{N,\nu}^{(L)}(\theta_k) \times P_L^N(\cos\chi)\cos\left[(\nu+N)(\phi_k-\gamma)\right] \quad (59)$$

In the general case given in equation (40), for each value of ν there will be 6 different $F_{N,\nu}^{(L)}$ leading to a total of 6(4n-3) unique complex valued $F_{N,\nu}^{(L)}(\theta_k)$ functions. When there are no chiral centers in a molecule, one can show that $F_{0,\nu}^{(L)} = (-1)^L F_{0,-\nu}^{(L)}$ so that with linearly polarized light, as assumed in equations (58) and (59), there will be $4(3n-2)$ unique $F_{N,\nu}^{(L)}$ functions, which in the case of $n=5$ leads to 52 unique functions. Depending on the transitions which contribute to the H_ν and on the structure of the molecule, many fewer than the maximum number of $F_{N,\nu}^{(L)}$ may be significant.

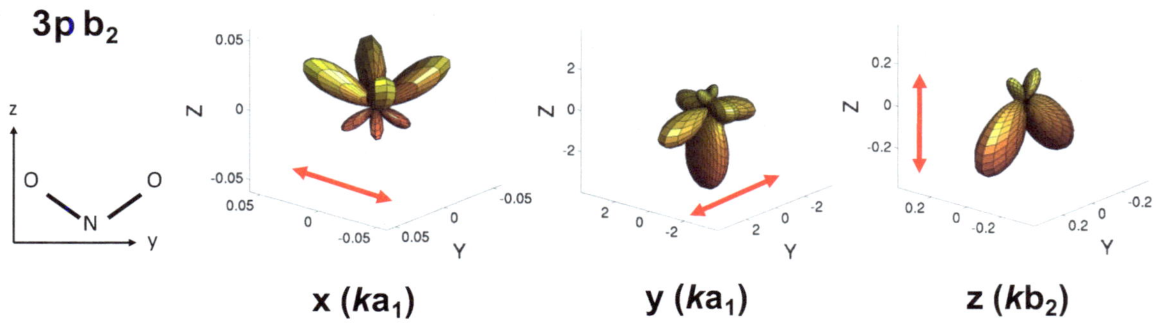

Figure 14: MFPADs computed for PI of the 3p b_2 Rydberg molecular orbital of NO_2 (state $5A'$, ground state geometry of NO_2) induced by light which is linearly polarized parallel to the x, y and z axes of the C_{2v} symmetry for a photoelectron energy of 0.5 eV. z is the main symmetry axis and yz corresponds to the molecular plane.

One model for the observed five photon DPI process in NO_2 assumes that ionization to the $NO_2^+(X\,^1A_1)$ ground state of NO_2 is induced by four photon absorption, followed by dissociation of the $NO_2^+(X\,^1A_1)$ by a fifth photon . Based on the similar recoil fragment distribution measured for ion-pair formation and dissociative photoionization, we assume for the illustration reported here that ionization takes place from the $5A'$ state of NO_2 whose dominant electronic structure is strongly mixed with the 3pσ (B_2) lowest member of the $R^*[(6a_1)^{-1}]$ Rydberg series converging to the $NO_2^+(X^1A_1)$ ground state [107]. In Fig. 14, we give computed MFPADs for ionization from the 3pσ (B_2) Rydberg state at the ground state geometry. This MFPAD, given by equation (23) is constructed from $F_{N,\nu}^{(L)}$ functions computed using SCFCHF type calculation.

To make a comparison with the experimental RFPAD data, we must then assume a model for the bound-to-bound part of the process. As an illustrative example, we give in Fig. **15** the measured and computed RFPADs for the five photon DPI of NO_2, assuming a [2,2,2,2] bound-to-bound scheme.

The measured RFPADs for the different DPI processes identified [109] display the general shape shown in Fig. **15** for a recoil velocity parallel to the polarization axis, with a remarkable electron emission anisotropy along the V_{NO^+} direction which favors electron emission in the direction of the recoil fragment: depending on the selected process, the dip observed in the V_{NO}^+ direction ($\theta_k \approx 0°$) is more or less pronounced.

On the right side of Fig. **15**, we present two different photoelectron angular distributions relative to the light polarization direction and the computed RFPAD for a recoil velocity parallel to the polarization axis. First, in red we give the expected angular distribution of single photon ionization from the 3p b_2 orbital, which has the form $\sigma\left[1+\beta P_2(\cos\theta_k)\right]$. We then consider two calculations where we assumed a [2,2,2,2] bound-to-bound type process as described above. In green we give the computed photoelectron angular distribution where we have integrated over all molecular orientations. Because this result is for a multiphoton process, the angular distribution can be more structured than in the one photon case, including terms up to $\cos^{2n}(\theta_k)$. Finally, in blue we give the computed RFPAD for this state and assumed bound-to-bound transitions. Although the agreement between the measured RFPAD on the left and the computed value on the right of Fig. **15** is not completely satisfactory, some dominant characteristics such as the strong emission anisotropy are well predicted. Further work is in progress exploring the sensitivity of final RFPAD to the parameters in the assumed model for the process, including the need to consider more geometries of the molecule, states contributing to the ionization, and the level of the treatment of the photoionization step.

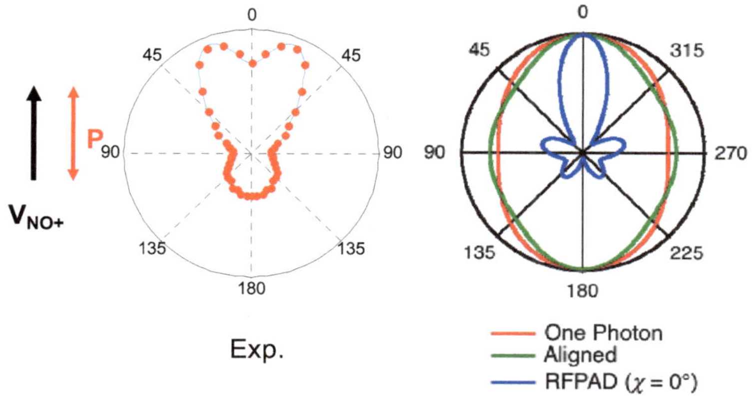

Figure 15: Measured RFPAD corresponding to a 5-photon DPI process for a NO^+ recoil fragment parallel to the axis of linearly polarized light, compared with three computed photoelectron angular distributions (see text): the computed RFPAD for the DPI process corresponding to PI of the 3p B_2 Rydberg state assuming a [2,2,2,2] bound-to-bound scheme is the blue line.

Another direction presently explored uses ultrashort XUV pulses to investigate MFPADs in resonant DPI of H_2 and D_2 [109] at the Attophysics group (SLIC).

X. CONCLUDING REMARKS

It has been well established in the last fifteen years that the measurement of molecular frame photoemission provides one of the most effective means to study the photoionization dynamics through the determination of the complex transition dipole elements which govern the process. Meanwhile, such studies increase our knowledge of the ionized molecular orbitals including the electronic correlation within the initial neutral state, of the correlation in the continuum state between the scattered and target electrons, and of the structure of the molecular ions. Many photoionization studies have been concerned with one-photon ionization of mainly diatomic and linear molecules, taking advantage of dissociative ionization and electron-ion momentum spectroscopy to record MFPADs. In such cases it was shown that analyzing the electron-ion emission patterns while relying on a unified formalism of the MFPADs induced by linearly then elliptically polarized light provides a very efficient mean for the extraction of the dipole matrix elements or dynamical parameters.

In this chapter, we have discussed the extension of this method to non-linear molecules and to multiphoton processes. These extensions are a work in progress. Despite the increasing complexity of a general description of the MFPADs and RFPADs, which may involve a significantly increased number of one dimensional $F_{N,\nu}^{(L)}(\theta_k)$ functions, different strategies can be developed to rationalize the expression of the RFPADs and disentangle the effects of photoionization from that of fragmentation dynamics or structural properties of the parent molecule. These studies strongly benefit from a detailed comparison of experimental and computed photoemission observables.

The angular characteristics of the photoelectron wave packets referred to the molecular frame are frequently identified as sensitive and efficient probes of interference and diffraction patterns in different contexts of molecular physics. The rapidly developing field of ultrafast electronic and nuclear dynamics in molecules involving tunnel ionization and/or weak field induced ionization, based on the use of forefront femtosecond and attosecond light sources and powerful imaging spectrometers, provides numerous illustrations. Complementing the investigation of dissociative phototionization reactions through electron-ion coincidence detection, the active alignment of molecules achieved through different schemes extends significantly the access to molecular frame photoemission, in particular in the ionization of larger polyatomic molecules, including chiral molecules.

ACKNOWLEDGEMENTS

The coworkers involved in the work discussed in this chapter, in particular M. Lebech, W.B. Li, J.C. Houver, C. Elkharrat, D. Toffoli, R. Carey, R. Montuoro, Y. J. Picard, P. Billaud, F. Catoire, L. Journel, R. Guillemin, M. Simon, are gratefully acknowledged. DD also acknowledges the contribution of T. Kachel (beamline UE 56 PGM-1 at BESSY), L. Nahon and coworkers (beamline DESIRS at SOLEIL) and C. Cornaggia, O. Gobert and coworkers (SLIC, Saclay), as well as the support through the BESSY IA-SFS program (Contract No. RII 3CT-2004-506008), the PICS France-USA CNRS contract and the Triangle de la Physique (DYNELEC No. 2008-046T). RRL acknowledges the support of the Robert A. Welch Foundation (Houston, Texas) under grant A-1020.

REFERENCES

[1] Dill, D. Fixed-molecule photoelectron angular distribution. *J. Chem. Phys.,* **1976**, *65*, 1130.

[2] Golovin, A.V.; Cherepkov, N.A., ; Kuznetsov, V.V. Photoionization of oriented molecules in a gas phase *Z. Phys., D* **1992**, *24*, 371.

[3] Shigemasa, E.; Adachi, J.; Oura, M.; Yagishita, A. Angular Distributions of 1sσ Photoelectrons from Fixed-in-Space N_2 Molecules. *Phys. Rev. Lett.,* **1995**, *74*, 359.

[4] Downie, P.; Powis, I. Molecule-Frame Photoelectron Angular Distributions from Oriented CF_3I Molecules. *Phys. Rev. Lett.,* **1999**, *82*, 2864.

[5] Eland, J.H.D.; Takahashi, M.; Hikosaka, Y. Photoelectron-fragment ion correlations and fixed-molecule photoelectron angular distributions from velocity imaging coincidence experiments *Faraday Discuss.*, **2000**, *115*, 119-126.

[6] Lafosse, A.; Lebech, M.; Brenot J.C. *et al.* Vector Correlations in Dissociative Photoionization of Diatomic Molecules in the VUV Range: Strong Anisotropies in Electron Emission from Spatially Oriented NO Molecules *Phys. Rev. Lett.*, **2000**, *84*, 5987.

[7] Hikosaka, Y.; Eland, J.H.D.; Watson, T.M.; Powis, I. Molecule frame photoelectron angular distributions from oriented methyl chloride and methyl fluoride molecules *J. Chem. Phys.,* **2001**, *115*, 4593.

[8] Gessner, O.; Hikosaka, Y.; Zimmermann, B. *et al.* U. $4\sigma^{-1}$ Inner Valence Photoionization Dynamics of NO Derived from Photoelectron-Photoion Angular Correlations *Phys. Rev. Lett.*, **2002**, *88*, 193002.

[9] Lucchese, R. R.; Lafosse, A.; Brenot, J. C. *et a.l* Polar and azimuthal dependence of the molecular frame photoelectron angular distributions of spatially oriented linear molecules *Phys. Rev. A* **2002**, *65*, 020702.

[10] Lafosse, A.; Brenot, J. C.; Guyon, P. M. *et al.* Vector correlations in dissociative photoionization of O_2 in the 20-28 eV range. II. Polar and azimuthal dependence of the molecular frame photoelectron angular distribution. *J. Chem. Phys.* **2002**, *117*, 8368-84.

[11] Lebech, M.; Houver, J. C.; Lafosse, A. *et al.* Complete description of linear molecule photoionization achieved by vector correlations using the light of a single circular polarization *J. Chem. Phys.,* **2003**, *118*, 9653-63.

[12] Lafosse, A.; Lebech, M.; Brenot, J.C. *et al.* Molecular frame photoelectron angular distributions in dissociative photoionization of H_2 in the region of the Q_1 and Q_2 doubly excited states. *J. Phys. B: At. Mol. Opt. Phys.*, **2003**, *36*, 4683.

[13] Dowek, D.; Lebech, M.; Houver, J.C.; R.R. Lucchese. Photoemission in the molecular frame using the vector correlation approach: from valence to inner-valence shell ionization *J. Electron Spectrosc. Relat. Phenom.*, **2004**, *141*, 211.

[14] Kugeler, O.; Prümper, G.; Hentges, R. *et al.* Intramolecular Electron Scattering and Electron Transfer Following Autoionization in Dissociating Molecules *Phys. Rev. Lett.*, **2004**, *93*, 033002.

[15] Lebech, M.; Houver, J.C.; Dowek, D.; Lucchese, R.R. Molecular Frame Photoelectron Emission in the Presence of Autoionizing Resonances. *Phys. Rev. Lett.*, **2006**, *96*, 073001.

[16] Dowek, D.; Pérez-Torres, J.-F.; Picard, Y. J. *et al.* Circular dichroism in photoionization of H_2. *Phys. Rev. Lett.*, **2010**, 104 233003.

[17] Heiser, F.; Gessner, O.; Viefhaus, J. *et al,* Demonstration of Strong Forward-Backward Asymmetry in the C1s Photoelectron Angular Distribution from Oriented CO Molecules. *Phys. Rev. Lett.*, **1997**, *79*, 2435.

[18] Cherepkov, N. A.; Raseev, G.; Adachi, J. *et al.* K-shell photoionization of CO: II. Determination of dipole matrix elements and phase differences. *J. Phys. B: At. Mol. Opt. Phys.*, **2000**, *33*, 4213–4236.

[19] Landers, A.; Weber, Th.; Ali, I. *et al.* Photoelectron Diffraction Mapping: Molecules Illuminated from Within. *Phys. Rev. Lett.*, **2001**, *87*, 013002.

[20] Jahnke, T.; Weber, Th.; Landers, A.L. *et al.* Circular Dichroism in K-Shell Ionization from Fixed-in-Space CO and N_2 Molecules. *Phys. Rev. Lett.*, **2002**, *88*, 073002.

[21] Motoki, S.; Adachi, J.; Ito, K. *et al.* Direct Probe of the Shape Resonance Mechanism in $2\sigma_g$-Shell Photoionization of the N_2 Molecule. *Phys. Rev. Lett.*, **2002**, *88*, 063003.

[22] Saito, N.; Ueda, K.; Fanis, A. D. *et al.* Molecular frame photoelectron angular distribution for oxygen 1s photoemission from CO_2 molecules. *J. Phys. B: At. Molec. Opt. Phys.* **2005**, *38*, L277-84.

[23] Adachi, J.; Ito, K.; Yoshii, H. *et al.* Site-specific photoemission dynamics of N_2O molecules probed by fixed-molecule core-level photoelectron angular distributions. *J. Phys. B: At. Mol. Opt. Phys.*, **2007**, *40*, 29–47.

[24] Li, W.B.; Montuoro, R.; Houver, J.C.; Haouas, A. *et al.* Photoemission in the NO molecular frame induced by soft-x-ray elliptically polarized light above the $N(1s)^{-1}$ and $O(1s)^{-1}$ ionization thresholds. *Phys. Rev. A,* **2007**, *75*, 052718.

[25] Fukuzawa, H.; Liu, X-J.; Montuoro, R. *et al.* Nitrogen K-shell photoelectron angular distribution from NO molecules in the molecular frame. *J. Phys. B: At. Mol. Opt. Phys.,* **2008**, *41*, 045102.

[26] Akoury, D.; Kreidi, K.; Jahnke, T. *et al.* The Simplest Double Slit: Interference and Entanglement in Double Photoionization of H_2. *Science*, **2007**, *318*, 949.

[27] Martin, F.; Fernandez, J.; Havermeier, T. *et al.* Single Photon–Induced Symmetry Breaking of H_2 Dissociation. *Science*, **2007**, *315*, 629.

[28] Rolles, D.; Braune, M.; Cvejanoviç, S. *et al.* Isotope-induced partial localization of core electrons in the homonuclear molecule N_2. *Nature*, **2005**, *437*, 711.

[29] Schöffler, M.; Titze, J.; Petridis, N. *et al.* R. Ultrafast Probing of Core Hole Localization in N_2. *Science*, **2008**, *320*, 920.

[30] Rolles, D.; Prümper, G.; Fukuzawa, H. *et al.* Molecular-Frame Angular Distributions of Resonant CO:C(1s) Auger Electrons. *Phys. Rev. Lett.*, **2008**, *101*, 263002.

[31] Liu, X.-J.; Fukuzawa, H.; Teranishi, T. *et al.* Breakdown of the Two-Step Model in *K*-Shell Photoemission and Subsequent Decay Probed by the Molecular-Frame Photoelectron Angular Distributions of CO_2. *Phys. Rev. Lett.*, **2008**, *101*, 083001.

[32] Sturm, F. P.; Schöffler, M.; Lee, S. *et al.* Photoelectron and Auger-electron angular distributions of fixed-in-space CO_2. *Phys. Rev. A.*, **2009**, *80*, 032506.

[33] Yamazaki, M.; Adachi, J.; Teramoto, T.; *et al.* 3D mapping of photoemission from a single oriented H_2O molecule. *J. Phys. B: At. Mol. Opt. Phys.*, **2009**, *42*, 051001.

[34] Lucchese, R.R. A simple model for molecular frame photoelectron angular distributions. *J. Electron Spectrosc. Relat. Phenom.*, **2004**, *141*, 201.

[35] Staudte, A.; Patchkovskii, S.; Pavicic, D. *et al.* Angular Tunneling Ionization Probability of Fixed-in-Space H_2 Molecules in Intense Laser Pulses. *Phys. Rev. Lett.*, **2009**, *102*, 033004.

[36] Allendorf, S.W.; Leahy, D.J.; Jacobs, D.C.; Zare, R.N. High-resolution angle-and energy resolved photoelectron spectroscopy of NO: Partial wave decomposition of the ionization continuum. *J. Chem. Phys.,* **2003**, *118*, 9653-63.

[37] Reid, K.L.; Leahy, D.J.; Zare, R.N. Complete description of molecular photoionization from circular dichroism of rotationally resolved photoelectron angular distributions. *Phys. Rev. Lett.*, **1992**, *68*, 3527.

[38] Tang, Y.; Suzuki, Y-I. ; Horio, T.; Suzuki, T. Molecular Frame Image Restoration and Partial Wave Analysis of Photoionization Dynamics of NO by Time-Energy Mapping of Photoelectron Angular Distribution. *Phys. Rev. Lett.*, **2010**, *104*, 073002.

[39] Stapelfeldt, H.; Seideman, T. Colloquium: Aligning molecules with strong laser pulses. *Rev. Mod. Phys.*, **2003**, *75*, 543.

[40] Ghafur, O.; Rouzée, A.; Gijsbertsen, A. *et al.* Impulsive orientation and alignment of quantum-state-selected NO molecules. *Nature Physics*, **2009**, *5*, 289-293.

[41] Thomann, I.; Lock, R.; Sharma, V. *et al.* Direct measurement of the angular dependence of the single-photon ionization of aligned N_2 and CO_2. *J. Phys. Chem.*, **2008**, *112*, 9382.

[42] Meckel, M.; Comtois, D.; Zeidler, D. *et al.* Laser-Induced Electron Tunneling and Diffraction. *Science*, **2008**, *320*, 1478-1482.

[43] Kumarappan, V.; Holmegaard, L.; Martiny, C. *et al.* Multiphoton Electron Angular Distributions from Laser-Aligned CS_2 Molecules. *Phys. Rev. Lett.*, **2008**, *100*, 093006.

[44] Akagi, H.; Otobe, T.; Staudte, A. *et al.* Laser Tunnel Ionization from Multiple Orbitals in HCl. *Science*, **2009**, *325*, 1364-1367.

[45] Holmegaard, L.; Hansen, J.L.; Kalhoj, L. *et al.* Photoelectron angular distributions from strong field ionization of oriented molecules. *Nature Physics*, **2010**, *6*, 428-432.

[46] Davies, J.A.; Continetti, R.E.; Chandler, D.W.; Hayden, C.C. Femtosecond Time-Resolved Photoelectron Angular Distributions Probed during Photodissociation of NO_2. *Phys. Rev. Lett.*, **2000**, *84*, 5983.

[47] Rijs, A.M.; Janssen, M.H.M., Chrysostom, E.T.H.; Hayden, C.C. *Phys. Rev. Lett.*, Femtosecond Coincidence Imaging of Multichannel Multiphoton Dynamics **2004**, *92* 123002.

[48] Gessner, O.; Lee, A.M.D.; Shaffer, J.P. *et al.* Femtosecond multidimensional imaging of a molecular dissociation. *Science*, **2006**, *311*, 219.

[49] Vredenborg, A.; Roeterdink, W.G.; Janssen, M.H.M. Femtosecond time-resolved photoelectron-photoion coincidence imaging of multiphoton multichannel photodynamics in NO_2 *J. Chem. Phys.*, **2008**, *128*, 204311.

[50] Bisgaard, C.Z.; Clarkin, O.J.; Wu, G. *et al.* Time-resolved molecular frame dynamics of fixed-in-space CS_2 molecules. *Science*, **2009**, *323*, 1464.

[51] Itatani, J. Levesque, J.; Zeidler, D. *et al.* Tomographic imaging of molecular orbitals *Nature*, **2004**, *432*, 867.

[52] Boutu, W.; Haessler, S.; Merdji, H. *et al.* Coherent control of attosecond emission from aligned molecules. *Nature Physics*, **2008**, *4*, 545.

[53] Smirnova, O.; Mairesse, Y.; Partchkovskii, S. *et al.* High harmonic interferometry of Multi-electron dynamics in molecules. *Nature*, **2009**, *460*, 972.

[54] Le, A-T.; Lucchese, R. R.; Lee, M. T. *et al.* Lin, C. D. Probing Molecular Frame Photoionization *via* Laser Generated High-Order Harmonics from Aligned Molecules. *Phys. Rev. Lett.*, **2009**, *102*, 203001.

[55] Dowek, D. Vector Correlations in Dissociative Photoionization of Simple Molecules Induced by Polarized Light. In: *Many Particle Quantum Dynamics in Atomic and Molecular Fragmentation*, Ullrich, J., Shevelko, V.P., Eds.; Springer: Berlin-Heidelberg, **2003**; pp.261-282.

[56] Lebech, M.; Houver, J.C.; Dowek, D. Ion-electron velocity vector correlations in dissociative photoionization of simple molecules using electrostatic lenses. *Rev. Sci. Instrum.* **2002**, *73*, 1866.

[57] Takahashi, M.; Cave, J.P.; Eland, J.H.D. Velocity imaging photoionization coincidence apparatus for the study of angular correlations between electrons and fragment ions. *Rev. Sci. Instrum.*, **2000**, *71*, 1337.

[58] Moshammer, R.; Fisher, D.; Kollmus, H. Recoil-Ion Momentum Spectroscopy and "Reaction Microscopes". In: *Many Particle Quantum Dynamics in Atomic and Molecular Fragmentation*, Ullrich, J., Shevelko, V.P. Eds.; Springer: Berlin-Heidelberg, **2003**; pp.33-58.

[59] Ullrich, J.; Moshammer, R.; Dorn, A. *et al.* Recoil-ion and electron momentum spectroscopy: reaction-microscopes. *Rep. Prog. Phys.* **2003**, *66*, 1463–1545.

[60] Zare, R. N. Dissociation of H2+ by Electron Impact: Calculated Angular Distribution *J. Chem. Phys.* **1967**, *47*, 204-15.

[61] Zare, R. N. Photoejection Dynamics. *Molec. Photochem.* **1972**, *4*, 1-37.

[62] Roentdek-Handels GmbH, Kelkheim, Germany.

[63] Détection: Temps, Position, Image (DTPI); LUMAT (FR2764 Univ Paris-Sud et CNRS), F-91405, Orsay, France.

[64] Vredenborg, A; Roeterdink, W. G.; and Janssen M. H. M. A photoelectron-photoion coincidence imaging apparatus for femtosecond time-resolved molecular dynamics with electron time-of-flight resolution of σ =18 ps and energy resolution $\Delta E/E$=3.5%. *Rev. Sci. Instrum.* **2008**, *79*, 063108.

[65] Weber, Th.; Czasch, A.; Jagutzki, O. *et al.* Fully Differential Cross Sections for Photo-Double-Ionization of D_2. *Phys. Rev. Lett.*, **2004**, *92*, 163001.

[66] Gisselbrecht, M.; Lavollée, M.; Huetz, A. *et al.* Photodouble Ionization Dynamics for Fixed-in-Space H_2. *Phys. Rev. Lett.*, **2006**, *96*, 153002.

[67] Born, M.; Wolf, E., *Principles of optics.* Seventh ed.; Cambridge University Press: Cambridge, 1999.

[68] Stert, V.; Radlo, W.; Schulz, C.P.; Hertel, I.V. Ultrafast photoelectron spectroscopy: Femtosecond pump-probe coincidence detection of ammonia cluster ions and electrons. *Eur. Phys. J. D* **1999**, *5*, 97-106

[69] Lucchese, R. R.; Raseev, G.; McKoy, V. Studies of differential and total photoionization cross sections of molecular nitrogen. *Phys. Rev. A* **1982**, *25*, 2572-87.

[70] Lucchese, R. R.; Carey, R.; Elkharrat, C.; Houver, J. C.; Dowek, D. Molecular frame and recoil frame angular distributions in dissociative photoionization of small molecules. *J. Phys.: Conf. Ser.* **2008**, *141*, 012009.

[71] Semenov, S. K.; Cherepkov, N. A. On the simplest presentation of the molecular frame photoelectron angular distributions for core levels. *J. Phys. B: At. Mol. Opt. Phys.*, **2009**, *42*, 085101.

[72] Lebech, M.; Houver, J. C.; Dowek, D.; Lucchese, R. R. Dissociative photoionization of N_2O in the region of the N_2O^+ ($B\ ^2\Pi$) state studied by ion-electron velocity vector correlation. *J. Chem. Phys.* **2004**, *120*, 8226-40.

[73] Toffoli, D.; Lucchese, R. R.; Lebech, M.; Houver, J. C.; Dowek, D. Molecular frame and recoil frame photoelectron angular distributions from dissociative photoionization of NO_2. *J. Chem. Phys.* **2007**, *126*, 054307.

[74] Lucchese, R. R.; Montuoro, R.; Grum-Grzhimailo, A. N. *et al.* Projection methods for the analysis of molecular-frame photoelectron angular distributions. I: Minimal parameterizations in the dipole limit. *J. Electron Spectrosc. Relat. Phenom.* **2007**, *155*, 95-9.

[75] Liu, X.-J.; Lucchese, R. R.; Grum-Grzhimailo, A. N. *et al.* Molecular-frame photoelectron and electron-frame photoion angular distributions and their interrelation. *J. Phys. B: At. Molec. Opt. Phys.* **2007**, *40*, 485-96.

[76] Prümper, G.; Rolles, D.; Fukuzawa, H. *et al* Measurements of molecular-frame Auger electron angular distributions at the CO C $1s^{-1}\ 2\pi^*$ resonance with high energy resolution. *J. Phys. B: At. Molec. Opt. Phys.* **2008**, *41*, 215101.

[77] Dixon, R. N. Recoil anisotropy following multiphoton dissociation *via* near-resonant intermediate states. *J. Chem. Phys.* **2005**, *122*, 194302.

[78] Lucchese, R. R., Molecular Photoionization. In Encyclopedia of Computational Chemistry, Schleyer, P. v. R., Ed. Wiley: Chichester, England, 2005.

[79] Lucchese, R. R.; Takatsuka, K.; McKoy, V., Applications of the Schwinger variational principle to electron-molecule collisions and molecular photoionization. *Phys. Rep.* **1986**, *131*, 147-221.

[80] Stratmann, R. E.; Zurales, R. W.; Lucchese, R. R., Multiplet-specific multichannel electron-correlation effects in the photoionization of NO. *J. Chem. Phys.* **1996**, *104*, 8989-9000.

[81] Stratmann, R. E.; Lucchese, R. R., A graphical unitary group approach to study multiplet specific multichannel electron correlation effects in the photoionization of O_2. *J. Chem. Phys.* **1995**, *102*, 8493-505.

[82] Orel, A. E.; Rescigno, T. N., Photoionization of ammonia. *Chem. Phys. Lett.* **1997**, *269*, 222-6.

[83] Rabadan, I.; Tennyson, J., R-matrix calculation of the bound and continuum states of the e-NO^+ sytem. *J. Phys. B: At. Molec. Opt. Phys.* **1996**, *29*, 3747-61.

[84] Bachau, H.; Cormier, E.; Decleva, P.; Hansen, J. E.; Martin, F., Applications of B-splines in atomic and molecular physics. *Rep. Prog. Phys.* **2001**, *64*, 1815-942.

[85] McCurdy, C. W., Jr.; Rescigno, T. N.; Yeager, D. L.; McKoy, V., The Equations of Motion Method: An Approach to the Dynamical Properties of Atoms and Molecules. In Methods of Electronic Structure Theory, Schaefer, H. F., III, Ed. Plenum Press: New York, **1977**, *3*, pp 339-86.

[86] Williams, G. R. J.; Langhoff, P. W., Photoabsorption in H_2O: Stieltjes-Tchebycheff calculations in the time-dependent Hartree-Fock approximation. *Chem. Phys. Lett.* **1979**, *60*, 201-7.

[87] Venuti, M.; Sterner, M.; Decleva, P., Valence photoionization of C_6H_6 by the B-spline one-centre expansion density functional method. *Chem. Phys.* **1998**, *234*, 95-109.

[88] Stener, M.; Decleva, P., Time-dependent density functional calculations of molecular photoionization cross sections: N_2 and PH_3. *J. Chem. Phys.* **2000**, *112*, 10871-9.

[89] Dehmer, J. L.; Dill, D., The continuum multiple-scattering approach to the electron-molecule scattering and molecular photoionization. In *Electron-Molecule and Photon-Molecule Collisions*, Rescigno, T.; McKoy, V.; Schneider, B., Eds. Plenum: New York, **1978**, pp 225-265.

[90] Miyabe, S.; McCurdy, C. W.; Orel, A. E.; Rescigno, T. N., Theoretical study of asymmetric molecular-frame photoelectron angular distributions for C 1s photoejection from CO_2. *Phys. Rev. A* **2009**, *79*, 053401.

[91] Vanroose, W.; Martin, F.; Rescigno, T. N.; McCurdy, C. W., Complete Photo-Induced Breakup of the H_2 Molecule as a Probe of Molecular Electron Correlation. *Science* **2005**, *310*, 1787-9.

[92] Tashiro, M., Application of the R-matrix method to photoionization of molecules. *J. Chem. Phys.* **2010**, *132*, 134306.

[93] Lin, C. D.; Le, A.-T.; Chen, Z.; Morishita, T.; Lucchese, R., Strong-field rescattering physics - self-imaging of a molecule by its own electrons. *J. Phys. B: At. Molec. Opt. Phys.* **2010**, *43*, 122001.

[94] Li, W. B.; Houver, J. C.; Haouas, A. *et al.* Photoemission in the molecular frame induced by soft X-ray elliptically polarized light. *J. Electron Spectrosc. Relat. Phenom.* **2007**, *156-158*, 30-37.

[95] Weiss, M.R.; Follath, R.; Sawhney, K.J.S. *et al.* The elliptically polarized undulator beamlines at BESSY II *Nucl. Instrum. Meth. in Phys. Res, A*, **2001**, *467-8*, 449-452.

[96] Schäfers, F.; Mertins, H.-Ch.; Gaupp, A. *et al.* Soft-x-ray polarimeter with multilayer optics: complete analysis of the polarization state of light. *Applied Optics*, **1999**, *38*, 4074-4088.

[97] Jahnke, T.; Weber, Th.; Osipov, T. *et al.* Multicoincidence studies of photo and Auger electrons from fixed-in-space molecules using the COLTRIMS technique. *J. Electron. Spectro. Relat. Phenom.*, **2004**, *141*, 229.

[98] Dowek, D.; Haouas, A.; Guillemin, R. *et al.* Recoil frame photoemission in inner-shell photoionization of small polyatomic molecules. *Eur. Phys. J. Special Topics* 2009, *169*, 85–93.

[99] http://www.synchrotron-soleil.fr/Recherche/LignesLumiere/DESIRS.

[100] Nahon, L.; Alcaraz, C. SU5: a calibrated variable-polarization synchrotron radiation beam line in the vacuum-ultraviolet range. *Applied Optics*, **2004**, *43*, 1024.

[101] Schmidt, V. Photoionization of atoms using synchrotron radiation. *Rep. Prog. Phys.* **1992**, 55, 1483-1659.

[102] Lörch, H.; Scherer, N.; Kerkau, T.; Schmidt, V. VUV light polarization measured by coincidence electron spectrometry. *J. Phys. B: At. Molec. Opt. Phys.* **1999**, *32*, L371-L379.

[103] Lebech, M.; Houver, J.C.; Dowek, D.; Lucchese, R.R. Dissociative Photoionization of N_2O in the region of the N_2O^+ ($C^2\Sigma^+$) state, studied by ion-electron velocity vector correlation. *J. Chem. Phys.* **2002**, *117*, 9248.

[104] Aitken, E. J.; Bahl, M. K.; Bomben, K. D. *et al,* Electron spectroscopic investigations of the influence of initial- and final-state effects on electronegativity. *J. Am. Chem. Soc.* **1980**, *102*, 4873-9.

[105] Herzberg, G., Molecular Spectra and Molecular Structure. III. Electronic Spectra and Electronic Structure of Polyatomic Molecules. Van Nostrand-Reinhold: New York, 1966.

[106] Wilkinson, L.; Whitaker, B.J. Some remarks on the photodynamics of NO_2. *Annu.Rep.Prog.Chem.Sect. C,* **2010**, *106*, 274-304.

[107] Elkharrat, C.; Picard, Y.J.; Billaud, P. *et al.* Ion Pair Formation in Multiphoton Excitation of NO_2 Using Linearly and Circularly Femtosecond Light Pulses: Kinetic Energy Distribution and Fragment Recoil Anisotropy. *J. Phys. Chem.* **2010**, *in press* DOI: 10.1021/jp103672h

[108] Werner, H.-J.; Knowles, P. J.; Lindh, R. *et al.* MOLPRO, a package of *ab initio* programs, 2002.6; Birmingham, UK, **2003**. see http://www.molpro.net.

[109] Pérez-Torres, J.F.; Morales, F.; Sanz-Vicario, J.L.;, Martin, F. Asymmetric electron angular distributions in resonant dissociative photoionization of H_2 with ultrashort XUV pulses. *Phys. Rev. A.*, **2009**, *80*, 011402(R).

CHAPTER 4

Cold Molecules, Photoassociation, Optical Pumping and Laser Cooling: The Cesium Case

Andrea Fioretti, Pierre Pillet* and Daniel Comparat

Laboratoire Aimé Cotton. CNRS. Université Paris-Sud. Bât. 505, 91405 Orsay cedex. France

Abstract: The cold-molecule field concerns the physics and the applications of molecular systems with translational temperature well below the 1 K range. The possibility of controlling all the motion as well as the internal quantum state of a sample of molecules is a long-term goal that opens the possibility for many new experiments and measurements ranging from fundamental constants to quantum chemistry and quantum computation. Although, direct laser cooling of the translational degree of freedom of molecules is still waiting for further technological or theoretical breakthroughs and ideas, many different techniques have proven to be successful in producing different types of cold as well as ultracold (*i.e.* T < 1 mK) molecules.

In this chapter, we will concentrate on the description of one of these techniques: the photoassociation of laser cooled atoms. We will report on the status of the art of this technique for the case of cesium atoms, describing all the main experimental findings. In particular, we will illustrate the different photoassociation schemes for molecule formation, the detection schemes through photoionization, the molecule trapping in a magnetic or dipolar trap, the vibrational cooling into a single vibrational state and finally the present prospects for rotational cooling.

1. INTRODUCTION

Full control of the dynamics of a quantum system is crucial in both physics and chemistry [1]. For atoms, precise control of both internal and external degrees of freedom has been achieved and has opened fascinating new fields [2]. Extension to molecules is not straightforward, but the impetus to prepare robust samples of trapped ultracold ground-state molecules with neither vibration nor rotation is strong. Indeed, significant advances are expected [3] in molecular spectroscopy, molecular clocks, fundamental tests of physics, super- or controlled photo-chemistry, and also in quantum computation based on polar molecules.

The temperature of the molecular sample is one of the key points because low temperature may allow for the long interrogation times necessary for high-precision measurements.

To achieve sub-Kelvin temperature and reach the now called "cold molecule" domain several techniques have been developed which can be classified under two main categories. The first one concerns cold molecules obtained starting with already formed molecules in general at room or higher temperature, which are subsequently subject to either a slowing or filtering or somehow cooling procedure capable to bring their translational temperature below the 1 K range. The second one concerns cold molecules formed starting with cold atom pairs and associating them into a molecule. These techniques have been recently reviewed in several books and articles [4-8].

In the following Table **1** we present an updated list (compared to the refs. [8-10]) of the cold molecular species with $T < 1$ K. Only the reference to the pioneering work done on each species is given. The recent combination of the He buffer gas pre-cooling with the Stark velocity filtering is extremely efficient and several cold molecular species have been produced using this promising technique: they are listed either under the "Cryogeny" or under the "Velocity filtering" category. The occurrence of some "exotic" molecules has been detected only as a loss process in a degenerate gas but not really produced and isolated. This is the case of Cs_3 and Cs_4 [11] and thus they do not appear in Table **1**. Similarly, several

***Address correspondence to Pierre Pillet:** Laboratoire Aime Cotton CNRS, Université Paris-Sud, bât 505, 91405 Orsay Cedex France; Tel: +33 (0) 1 6935 2005; E-mail: Pierre.Pillet@lac.u-psud.fr

Gennadi Ogurtsov and Danielle Dowek (Eds)

photoassociation experiments have produced electronically excited cold molecules (as single example, Yb [12]) but we do not include them in the list as they have a lifetime limited to some tens of nanoseconds. Furthermore, we do not always specify the isotope difference, for instance NH_3 can be the same as $^{15}NH_2D$. Finally, cold molecule studies in NanoDroplet are not listed in Table **1**, a short list is given in ref.[13] and this has been reviewed in refs.[14-18].

In this chapter we will concentrate on the description of one of these techniques: the *photoassociation* of laser cooled atoms. We will report on the status of the art of this technique for the case of cesium atoms, describing all the main experimental findings. In particular, we will illustrate the different photoassociation schemes for molecule formation, the detection schemes through photoionization, the molecule trapping in either a magnetic or dipolar trap, the vibrational cooling into a single vibrational state and finally the present prospects for rotational laser cooling.

Table 1. Slow and cold molecule list [4, 8, 19-104].

Method	Molecule	T(MK)	N
Feshbach, RF [19, 20]	$^{85,87}Rb_2$ [21-23], Cs_2 [24], $^{40}K_2$ [25], Li_2 [26-28], Na_2 [29], ^{40}KRb [30], $^{41}K^{87}Rb$ [31], Cr_2 [32], Li_3 [33]	0.1	10^5
Photoassociation [34]	Cs_2 [35], H_2 [36], Rb_2 [37], Li_2 [38], Na_2 [39], K_2 [25, 40, 41], He_2^*, [42], Ca_2 [43], KRb [44], RbCs [45], NaCs [46], LiCs [47]	100	$2\ 10^5$
Three body collision	Rb_2 [37] Li_2 [26, 48]	0.2	$2\ 10^6$
Cryogeny	CO [49], VO [50] CaF [51], PbO [52, 53], O_2 [54] NH [55], ND, CrH, MnH [56], ND_3, H_2CO [57], YbF [58], NH, NH_3, O_2, ThO, Naphthalene [59]	$4\ 10^5$	10^{12}
Field Slowing: Stark [60] Rydberg Optical Zeemann	CO [61], NH_3, ND_3 [62, 63], OH [64, 65], OD [66], H_2CO [67], NH [68], SO_2 [69], C_7H_6N [70], C_6H_7NO [70], YbF [71], LiH [72], CaF [73] H_2 [74-77] C_6H_6 [78], NO [79] O_2 [80]	10^4	10^6
Beam collision Beam dissociation	NO [81], KBr (13K) [82], ND_3 [83] NO [84]	$4\ 10^5$ $1.6\ 10^6$	
Rotating Nozzle	O_2, CH_3F,SF_6 [85, 86], CHF_3 [87], perfluorinated C_{60} [88]	10^6	
Velocity filtering	H_2CO [67], ND_3 [89], D_2O [90], CH_3F [91], CH_3CN [92], H_2O, D_2O, HDO [93], NH_3, CH_3I, C_6H_5CN, C_6H_5C1 [94]	10^6	10^9
Sympathetic cooling [4, 8, 95]	BeH+, YbH+ [96], AF350$^+$=$C_{16}H_{14}N_2O_9S^+$[97], MgH^+ [08], O_2^+, MgO+, CaO+ [99], H_2^+, H_3^+ [100], BaO^+ [101], NeH+,N_2^+, OH^+, H_2O^+, HO_2^+,ArH^+, CO_2^+, KrH^+, C_4F_8, $R6G^+$ [8, 102], Cyt^{12+}, Cyt^{17}+ [103], GAH^+ = $C_{30}H_{46}O_4^+$ [104]	$2\ 10^4$	10^3

This chapter collects the main results and ideas of the experimental and theoretical work on cold molecules performed in our group at the Laboratoire Aimé Cotton, Orsay, during the last years. This work has already been presented in several articles that are cited underway. For a detailed description of each result the reader can refer to the specific article.

2. FORMATION AND DETECTION OF Cs_2 MOLECULES

Among the current methods used to prepare dense and ultracold samples of molecules, we report here on photoassociation (PA) of laser cooled Cs atoms. In particular, we demonstrate efficient production of deeply bound Cesium dimers and vibrational cooling in the v=0 level of the X $^1\Sigma_g^+$ ground electronic singlet state. In Cesium, the PA process corresponds to the reaction

$$Cs(6s,F) + Cs(6s,F) + h\nu_L \rightarrow Cs_2^*(\Omega(6s+6p);v,J) \quad (1)$$

Two colliding atoms in a F hyperfine level of their ground state 6s absorb one laser photon at frequency $h\nu_L$ red-detuned from the atomic resonance frequency ($6s + 6p_j$, j = 1/2 or 3/2) and form a molecule in a well defined rovibrational level (v, J) of an excited molecular state Ω correlated to one of the asymptotes ($6s+6p_{1/2}$) or ($6s + 6p_{3/2}$). The resolution of the PA process is limited by the width ($\sim k_BT$) of the statistical distribution of the relative kinetic energy of the colliding atoms. For ultracold atoms, ($k_BT \sim h$ x 2 MHz at T ~ 100 μK), it is smaller than most of other relevant energy spacings of the system. Thus, cold atom PA is a powerful tool for high-resolution molecular spectroscopy. It has given access to previously unexplored regions of molecular potential curves at distances much larger than those of well-known chemical bonds. PA spectroscopy has been reported first for homonuclear alkali dimers (Na_2 [105] and Rb_2 [106]) and then for other homonuclear as well as heteronuclear dimers, involving most of the elements that have been successfully laser cooled [107]. In general, photoassociation is detected as a "trap loss", *i.e.* cold atoms that undergo PA gain enough kinetic energy to leave the traps where they are stored. This happens after some either radative decay or (auto/photo)-ionization process or also some predissociation process takes place. In some cases, radiative stabilization into ground state dimers is possible, as predicted in ref. [108] and actually observed in our group for the first time [35].

The description of the wealth of spectroscopical results obtained by photoassociation studies on the Cesium dimer by our group and others in the last ten years is beyond the scope of this chapter. It concerns mainly the ground molecular states dissociating into the 6s + 6s asymptote, the first excited states dissociating into the $6s + 6p_{1/2,3/2}$ asymptotes and some other highly excited states. These results can be found in refs. [109-112] for the ground states, in refs. [113, 114] for the $6s + 6p_{1/2}$ excited states, in refs. [115-124] for the $6s + 6p_{3/2}$ excited states and finally in refs. [125, 126] for other excited states.

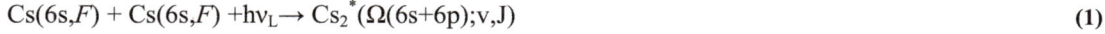

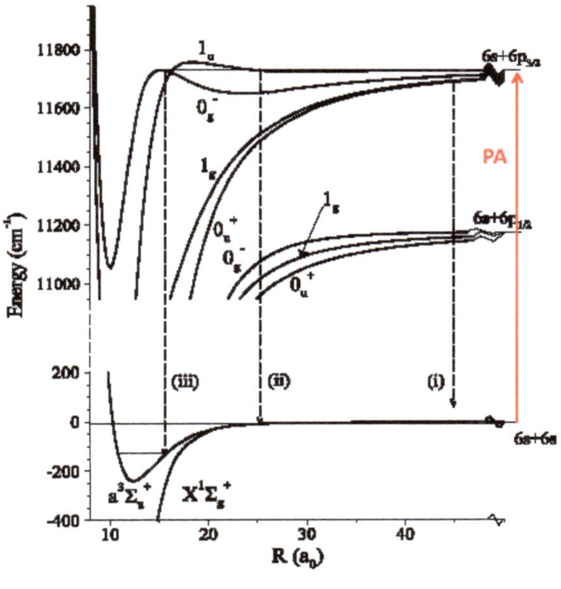

Figure 1: Long-range Cs_2 attractive Hund's c case molecular potential curves correlated to the (6s + 6s) and ($6s + 6p_{1/2,3/2}$) dissociation limits. The attractive 0_u^- and 2_u states, forbidden for dipolar transition, are not shown. Long range excitation by the PA laser is represented by the red arrow. The following spontaneous emission, depicted by dashed arrows, leads respectively to the formation of cold molecules in cases (ii) and (iii), and to dissociation into two free atoms in case (i). The Figure is extracted from ref. [127].

The PA process is depicted in Fig. **1**. A pair of identical ground state atoms interact at large interactomic distances R through their R^{-6} Van der Waals interaction while for a homonuclear excited atom pair the R^{-3} dipole-dipole interaction is dominant. Vibrational levels with a very large elongation (from few tens up to few hundreds of atomic units) are then efficiently populated by photoassociation. To produce ground state cold molecules, from a classical point of view, the ideal vibrational motion of the photoassociated molecule should slow down in the short to intermediate distance range (say, around 5-20 atomic units) to let spontaneous decay occur before going back towards the long distance range, where decay simply produce a pair of more energetic atoms. This was observed in ref. [35] by using the 0_u^- and 1_u double-well potential curves correlated to the Cs(6s)+Cs(6$p_{3/2}$) limit, shown in Fig. **1**. These peculiar states, known as pure long-range states [128], result from the competition of spin-orbit interaction and long-range dipole-dipole interaction. The low R –variation of the left edge of the outer potential well induces a "speed bump" in the distance range appropriate for the radiative decay into stable vibrational levels of the ground state electronic potentials.

Within their short lifetime of a few tens of nanoseconds, the electronically excited molecules created by photoassociation most often spontaneously decay back to a pair of "hot" atoms, *i.e.* atoms with large relative kinetic energy. In magneto-optical traps, the fluorescence intensity decrease due to the escape of the hot atoms from the trap provides a simple detection method. Nevertheless, the detection of cold dimers, when allowed, provides an alternative and very efficient way of detecting the PA process.

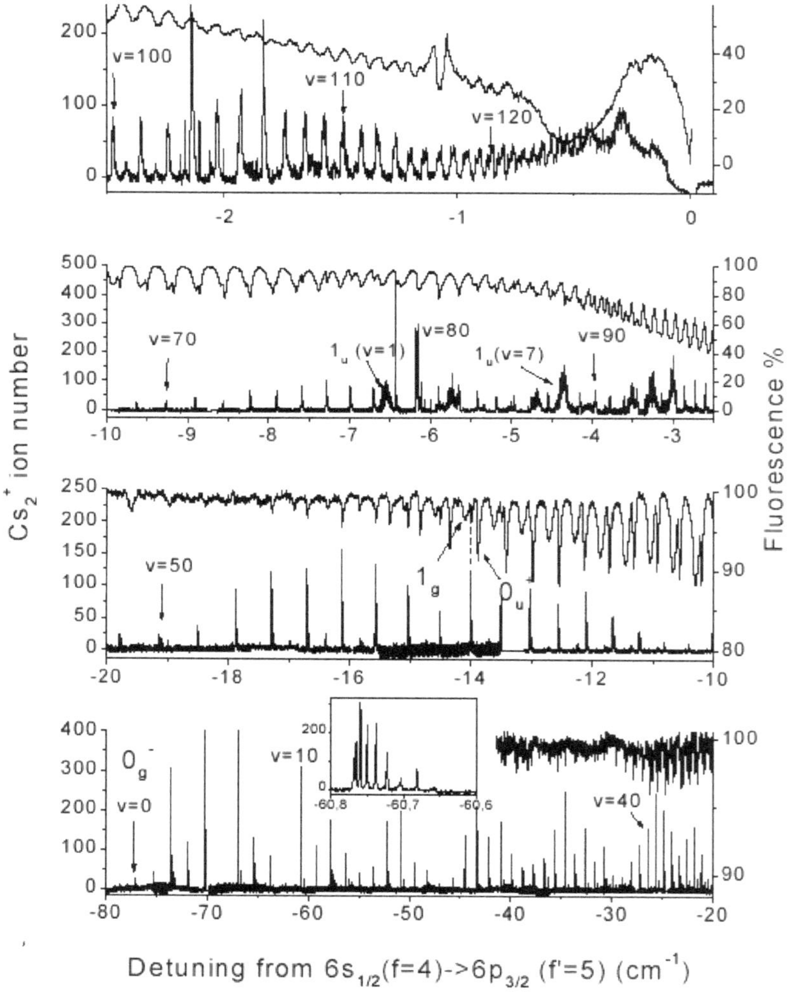

Figure 2: Cs_2^+ ion signal (lower curve) and trap fluorescence yield (upper curve) versus detuning of the PA laser relative to the 6s + $6p_{3/2}$ dissociation limit. The rotational progression of the v = 10 level of the 0_g^- long-range potential well is shown in the inset. The dashed line indicates the correspondence of a vibrational level of the 0_g^- state on both spectra. This Figure is extracted from ref. [127].

2.1. Molecules in the $a^3\Sigma_u^+$ Triplet Ground State

Fig. **2** displays a typical PA spectrum from a Cs MOT, when the PA laser frequency (in our setup a cw Ti:Sa laser, intensity 300 Wcm^{-2}, pumped by an Argon-ion laser) is scanned below the 6s + $6p_{3/2}$ dissociation limit. The upper trace corresponds to the fluorescence spectrum due to trap loss, revealing rovibrational progressions for all attractive potentials which can be reached by photoassociation; in this case they are the 0_g^-, the 0_u^+, the 1_g and the 1_u states in the Hund's "c" case nomenclature (cf. Fig. **1**). Both decays into a pair of hot atoms and into a stable molecule, induce a decrease of the trap fluorescence. The lower trace is the Cs_2^+ yield obtained by photoionization of the stable Cs_2 cold molecules created either in the ground state (trough 1_u excitation) or in the lowest triplet $a^3\Sigma_u^+$ state (through 0_g^- excitation). The electronically excited molecules formed by photoassociation have too short lifetime to give a significant contribution to the photoionization signal. The 0_g^- (6s + $6p_{3/2}$) and 1_u(6s + $6p_{3/2}$) double-well potentials are the only states which contribute significantly to the molecular ion signal [115, 117]. In the former case, the cold molecules are distributed over several rovibrational levels with binding energies in the middle of the lowest triplet state $a^3\Sigma_u^+$, while in the latter case, the populated levels are close to the dissociation limit, where the gerade (*g*) and ungerade (*u*) characters are no longer good symmetries. The Cs_2^+ ion spectrum in Fig. **2** exhibits 133 well resolved lines assigned to the vibrational progression of the 0_g^-(6s + $6p_{3/2}$) outer well, starting at v = 0. The rotational structure, shown for v = 10 in the inset, is observed up to J = 8 for most of the vibrational levels below v = 74. The double-well route *via* the 0_g^- potential correlated to the 6s+$6p_{3/2}$ was also demonstrated for Rubidium [37]. For other alkalis the 0_g^- outer well is still present, but located too far out to provide a Condon point in the desired distance range, and no significant formation of ultracold molecules can occur. The double-well route is an optimized compromise between an efficient photoassociation at long distances and a quite reasonable branching ratio for spontaneous decay at intermediate distances.

2.2. Molecule Detection by Resonance Enhanced MultiPhoton Ionization

The detection of cold molecules is made through REMPI (Resonant Enhanced MultiPhoton Ionization) detection [115], whose sensitivity can reach one single ion [125]. The cold molecules are ionized into Cs_2^+ molecular ions by a pulsed dye laser (7 ns duration, 1 mJ energy, focusing spot ~ $1mm^2$, 10 GHz resolution) pumped by the second harmonic of a Nd-YAG laser at 10 Hz repetition rate. The photoionization process typically involves molecules produced during the last 10 ms before the laser pulse.

Top of Fig. **3** shows the REMPI spectrum recorded by scanning the frequency of the pulsed REMPI laser in the range 13500 — 14500 cm^{-1} with the PA laser tuned to the v = 79 vibrational level of the 0_g^-(6s + $6p_{3/2}$) state. In this case cold molecules are formed mainly in the ground triplet state. The REMPI process consists in a first resonant step corresponding to a transition between one ro-vibrational level of the triplet $a^3\Sigma_u$ state and a level of either the $(2)^3\Pi_g$ or the $(2)^3\Sigma_g$ state converging to the dissociation limit 6*s* + *5d.* The second step corresponds to one-photon ionization of this excited intermediate rovibrational level into the Cs_2^+ continuum. After the photoionization laser pulse, a pulsed high electric field (3kV/cm, 0.5 μs) is applied at the trap position by a pair of grids spaced by 15 mm. Ions expelled from the association region cross a 6 cm free field zone that acts as a time-of-flight mass spectrometer capable to separate the Cs_2^+ from spurious Cs^+ ions. Both ionic species are separately detected by a pair of microchannel plates and the signals are recorded by a boxcar integrator.

The total efficiency of the process is limited by the ion recollection rate (80 %) and the microchannel plate efficiency (35 %). The REMPI efficiency is roughly estimated, in this case, to be around 10 % by comparing the Cs_2^+ ion signal with the trap loss fluorescence signal and by using calculated branching ratios between bound-bound and bound-free transitions for the photoassociated molecules. One of the most interesting features of the REMPI tool for the molecule detection is the fact that by changing the frequency and the bandwidth of the REMPI laser, the detection process can be made either *non selective* (large bandwidth laser, very spectrally dense intermediate state) or, on the contrary, *extremely selective* (narrower bandwidth laser, spectrally clean intermediate state). In the first case we will detect most of the molecules produced by the PA process irrespective of their ro-vibrational level while in the second case we are able, through a precise spectroscopic knowledge of the involved molecular states, to monitor the population of molecules produced in a specific vibrational level. In the case shown in Fig. **3**, we have used a selective narrow bandwidth laser and observed a quite clean spectrum with distinct features meaning that only a few initially populated rovibrational levels, for which the two-photon process is resonant, are efficiently ionized. For frequencies below 13850 cm^{-1}, the scan yields a well resolved spectrum (bottom of Fig. **3**). The first resonant step of the REMPI process corresponds closely to transitions between rovibrational levels of the lowest triplet state $a^3\Sigma_u^+$ and the $(2)^3\Sigma_g^+$ state converging to the dissociation limit 6s + 5d. The analysis presented in ref. [126] shows that vibrational levels between v = 16 and 24 of the lowest triplet $a^3\Sigma_u^+$ state are populated. In general, cold molecules produced by PA in the lowest triplet state are always vibrationally excited. A comprehensive analysis of all the possible schemes used until now to efficiently detect cold molecules in the $a^3\Sigma_u^+$ triplet ground state is reported in ref.[126]. In this reference it is shown that triplet molecules are efficiently detected by ionizing them through any of the following states: the $(1)^3\Sigma_g^+$ state in the region of 8500 cm^{-1}, the $(2)^3\Sigma_g^+$ state, in the region of 13000-14000 cm^{-1}, the $(3)^3\Sigma_g^+$ state in the region of 15800-16100 cm^{-1} and, finally, the $(2)^3\Pi_g^+$ state, in the region of 13800-14200 cm^{-1}. The rather large vibrational structure and the absence of the spin-orbit interaction of the states of Σ type allows selective detection of individual vibrational levels in the ground state (see Fig. **3b**). On the contrary, the complex structure of the Π intermediate state, discussed in ref.[125], allows non-selective detection of ground state triplet molecules. The photoionization region around 16000 cm^{-1} is of particular interest because it allows simultaneous selective detection of triplet molecules as well as deeply bound singlet molecules, as discussed in Section 2.4.

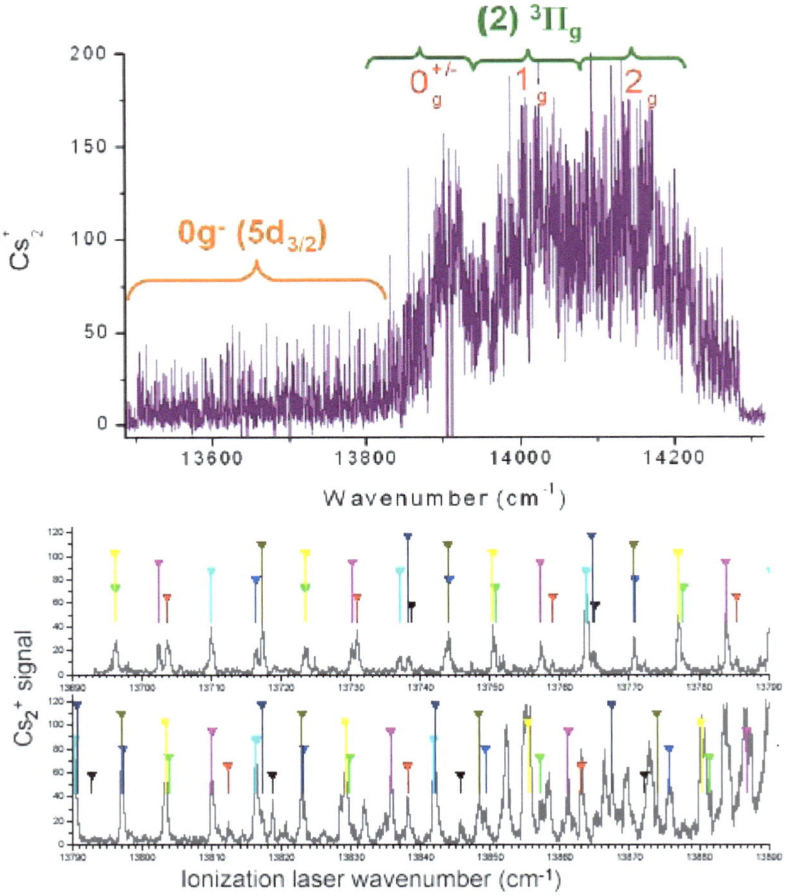

Figure 3: Photoionization spectrum of Cs_2 cold molecules formed after spontaneous emission from the v = 79 vibrational level of the $0_g^-(6s + 6p_{3/2})$ state. Each transition is indicated by a triangle. The Figure is extracted from ref. [127].

2.3. Raman Photoassociation

Photoassociation schemes are not restricted to the interaction of one light field with the atomic sample, but more sophisticated schemes with multiple photon absorption and emission can be considered to access other molecular states. In the case of two-color PA a photon from each field can be absorbed, leading to a highly excited molecular state, or the absorption of a second photon,

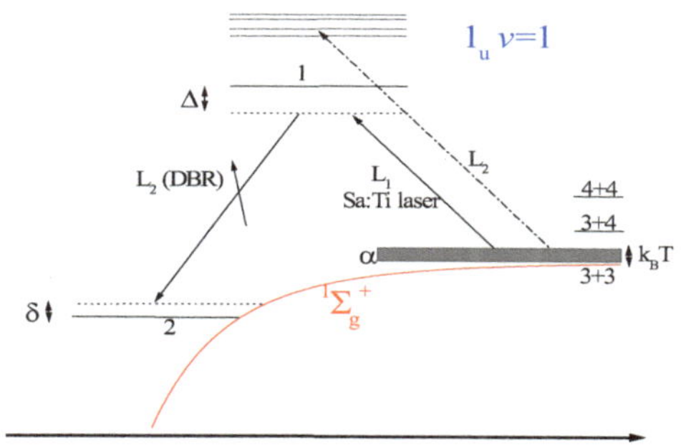

Figure 4: Scheme of Raman photoassociation. Δ is the detuning of the laser field L_1 with respect to the level of the intermediate state(a given state of 1_u (6s + $6p_{3/2}$) potential here). δ is the detuning with respect to the level of the final state. The grey shaded area above the lowest hyperfine asymptote 3 + 3 indicates the thermal distribution of atoms in the MOT.

leading mainly directly to the ground molecular state. When the first photon is out of resonance we can use the term "Raman photoassociation". This approach can be used to enhance the formation of cold molecules, for high-resolution spectroscopy or to study or modify the scattering properties of the colliding atoms.

We used this scheme in the cesium case for accurate spectroscopic investigation of weakly bound molecular levels in the ground state, to perform lineshape analysis and to derive molecular constants like the Van der Waals long-range dispersion coefficient C_6 and the exchange term. The results are reported in Refs. [109, 110]. The scheme of Raman PA spectroscopy is shown in Fig. **4**. A first laser L_1 is tuned close to an excited molecular level. The frequency of a second laser L_2 is scanned to probe resonances in the ground state The L_2 laser could also excite directly another excited level, but care is taken to the choice of the intermediate excited level to make this process negligible.

Cesium has nuclear spin I = 7/2, leading to a large hyperfine structures in the ground (roughly 9.2 GHz) as well as in the excited atomic state. The molecular states of cesium dimer have a hyperfine structure that gives rise to a multitude of vibrational energy levels, which complicates the spectra especially close to dissociation. The use of the dark SPOT technique [129], leading to a trapped sample in the lower hyperfine state, allowed us to consider only photoassociation from the lowest hyperfine F = 3 + F = 3 asymptote.

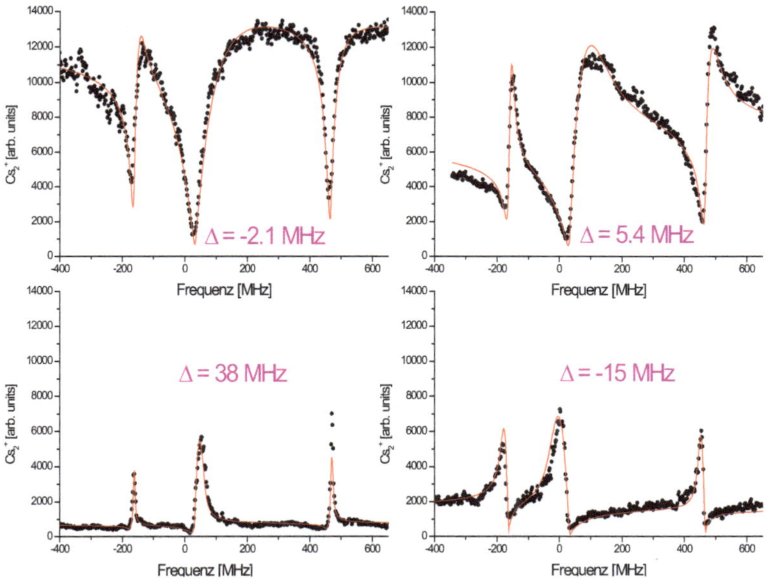

Figure 5: Rotational series of a vibrational level 0.6 cm^{-1} below the hyperfine asymptote 3 + 3 measured by two-color PA for 4 different Raman detunings. The progression corresponds to the values l = 0, 2, 4 of the molecular rotation of the nuclei (see [110] for details).

A sample of line shape fits is reported here as well as the investigation of detuning dependent line shifts (see Fig. **5**). Calculations according to the model of ref. [130] and measurements are in very good agreement for a wide range of experimental conditions even in the case of nearly blending rotational lines. It was possible to observe and reproduce the transition from almost resonant "frustrated" photoassociation *via* an intermediate level to a far detuned regime (see Fig. **5**). In the former case, the PA produced by the fist laser is "frustrated" by the second one, which displaces the excited level when the resonance condition with a ground state level is attained. In this case, dips appear instead of sharp peaks. Care has to be taken also to the interpretation of far detuned regime, where the observed sharp peak of Cs_2^+ ions is not due to selective population of a single well defined molecular level but still due to spontaneous emission losses from the intermediate state. These Fano's type of line profiles contain also information about the energy distribution of atoms in the cloud. Thus an investigation of the line profiles versus the detuning parameters δ, Δ can be used to derive collision properties.

In our works [109, 110] more than 100 high lying level energies of the lowest electronic states in Cs_2 have been observed. From this analysis we have determined the long-range dispersion coefficient C_6 and we have obtained the first experimental determination of the amplitude of the asymptotic exchange term.

2.4. Molecules in the X $^1\Sigma_g^+$ Singlet Ground State

The previous mechanism, based on PA towards long- range states with favorable decay at intermediate distances, relies strongly on the specific shape of the molecular potential curves. It seems by now limited to Rb_2 and Cs_2 for the 0_g^- symmetry, and to Cs_2 for the 1_u symmetry. This gives access, in the cesium case, mainly to ground state molecules in the triplet $a^3\Sigma_u^+$ manifold or to very weakly bound molecules near the dissociation limit, where the hyperfine structure mixes the u and g symmetries.

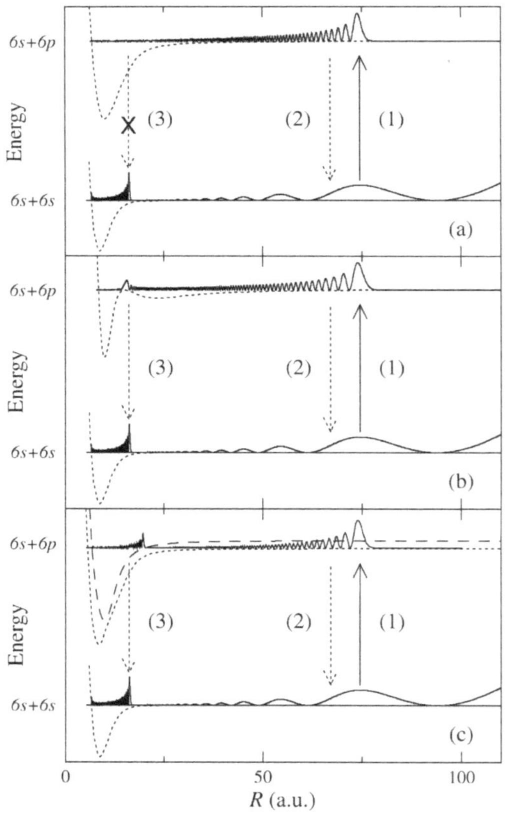

Figure 6: Photoassociation from the cesium 6s+6s continuum [reaction (1)] to (a) a typical potential well; (b) a double-well potential (e.g. $0_g^-(6s+6p_{3/2})$); (c) two coupled states (e.g. $0_u^+(6s+6p_{1/2,3/2})$). The system decays by spontaneous emission either back to the continuum [reaction (2)] or to a bound level of the ground state [reaction (3)]. For the case (a), reaction (3) is usually unlikely, This figure is taken from ref. [113].

To access the singlet X state, in particular, its deeply bound levels, it is necessary to rely on a different formation mechanism. This possibility is provided by the more general pattern of interactions in diatomic molecules taking place when two (or more) molecular states are coupled by some interaction (like spin-orbit, rotational or other) at a given internuclear distance. In this case, vibrational levels of the different electronic states that are quasi-degenerate in energy can be strongly mixed and real vibrational wave functions are superpositions of the eigenfunctions of the separate potential curves. An example of such a situation is given by the $0_u^+(6s + 6p_{1/2})$ and $0_u^+(6s + 6p_{3/2})$ states of Cs_2, coupled by the spin-orbit interaction. A detailed analysis of this mechanism, as performed in ref. [113]. suggests that the resonant coupling between vibrational levels of the $0_u^+(6s + 6p_{1/2})$ and 0_u^+ $(6s + 6p_{3/2})$ states enables simultaneously efficient photoassociation at large distance and reasonable formation rate of ultracold molecules

This process is depicted in Fig. **6**. The PA excites the 0_u^+ $(6s + 6p_{1/2})$ component of the vibrational wave function at large distance while the coupling transfers population to the 0_u^+ $(6s + 6p_{3/2})$ component of the wave function at short distances, favoring spontaneous emission towards stable bound levels of the $X^1\Sigma_g^+$ Cs_2 ground state.

Although in the experiment of ref. [113] the vibrational distribution in the X $^1\Sigma_g^+$ ground state was not accessible, the numerical calculation showed that most of the produced molecules were expected to be in vibrational levels greater that v = 128, meaning that they are bound by less than 20 cm^{-1}. Nevertheless, a major goal of these experiments is to produce molecules that are also vibrationally cold, *i.e.* deeply bound.

A determinant step towards this objective has been possible by changing the detection strategy as follows. As the PA transitions leading to a favorable decay into deeply bound levels of the ground X $^1\Sigma_g^+$ state were not known in advance, the idea was to set up a detection scheme that could efficiently ionize most of the

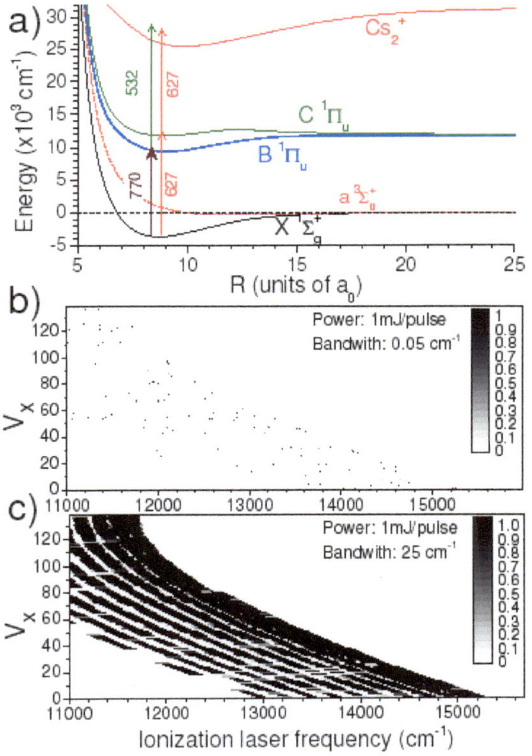

Figure 7: REMPI detection scheme of deeply-bound ground state Cs_2 molecules with a broadband laser at 770 nm (with an extra laser at 532 nm) *via* the $B^1\Pi_u$ state, and with a narrowband laser at 627 nm *via* the $C^1\Pi_u$ state. Transition probabilities (in levels of grey) of the ground-state vibrational levels v_X toward levels of the B state, as functions of their energy difference $E_{X\text{-}B}$ for a laser linewidth of 0.05cm^{-1} (b) and of 25cm^{-1} (c), with identical power (1 mJ/pulse). The probability is put to unity for a saturated transition. This Figure is taken from ref. [131].

low lying, v < 35, vibrational levels of this state. This is in fact possible by choosing the B $^1\Pi_u$ as intermediate state for photoionization. This two-photon, two-color process is represented by the dark red and grey arrows in Fig. **7**. A broadband (FHWM ~ 25 cm^{-1}) dye (LDS 751) laser, pumped by the second harmonic of a pulsed Nd:YAG laser, excites the molecules on the vibrational transitions X $^1\Sigma_g^+$ (v) → B $^1\Pi_u$ (v'). The second harmonic of the pulsed Nd:YAG laser ionizes the so-excited molecules and Cs_2^+ ions are detected through a time-of-flight mass spectrometer. Such broadband laser, tunable in a wide range (13500 – 14500 cm^{-1}), is able to detect cold molecules in a broad distribution of rovibrational levels. Then by scanning the PA cw Ti:Sa laser, we have found several new pathways for the formation of cold molecules in rhe singlet ground state. Once determined a particular PA frequency producing singlet molecules, we have analyzed the so-formed vibrational distribution in the ground state by changing again the photoionization scheme. We switched back to a selective REMPI scheme, performed with a narrower band laser (FWHM 0.3 cm^{-1}), proceding through the C $^1\Pi_u$ intermediate state. This two-photon process at 627 nm is represented by the bright red arrows in Fig. **7**. The results, reported in ref. [131], showed that low-lying levels v = 1 — 12 of the X $^1\Sigma_g^+$ ground state could be efficiently populated. Almost no molecules are fiund in the vibrational level v = 0. The formation mechanism for these molecules has been discussed in detail in ref. [131] and is shown in Fig. **8**.

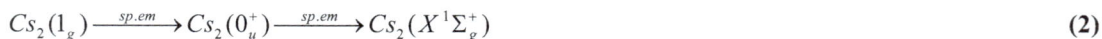

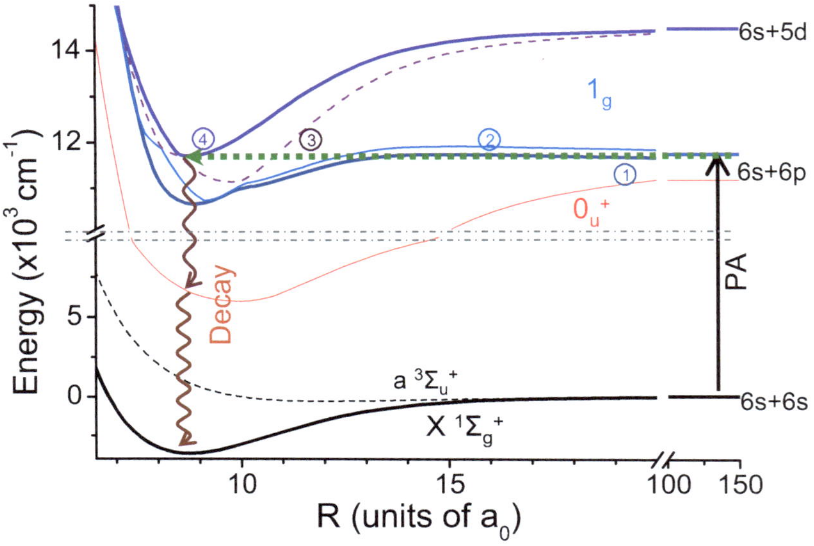

Figure 8: Theoretical Cs_2 molecular potential curves including spin-orbit interaction [132, 133] relevant to the present PA and cold molecule formation process. The PA laser excites levels of a long-range 1_g curve (label 1), which is coupled to the v = 0 level of the short-range 1_g curve (label 4) through several avoided crossings involving 1_g curves (labeled 2 and 3). Formation of deeply-bound ground state molecule proceed through a spontaneous emission cascade *via* the 0_u^+ states. This Figure is taken from ref. [131]

It involves a photoassociation pathway towards a level that is resonantly coupled to a more internal one plus a two-photon decay step. The process is

$$Cs_2(1_g) \xrightarrow{sp.em} Cs_2(0_u^+) \xrightarrow{sp.em} Cs_2(X\,^1\Sigma_g^+) \qquad \textbf{(2)}$$

The cw PA laser is tuned ~ 1 cm^{-1} on the red of the atomic transition $6s_{1/2}$ —> $6p_{3/2}$ and excites a vibrational level (v, J) of a state, spectroscopically labelled 1_g, converging towards the electronically excited limit $6s_{1/2} + 6p_{3/2}$. As this electronic state is coupled to another state of 1_g symmetry asymptotically connected with the (6s + 5d) dissociation limit, some vibrational levels, quasi-degenerate in energy with the v = 0 of the inner state, are mixed and a substantial fraction of the population is transferred at short interatomic distance. Here, although the so-photoassociated molecules can spontaneously decay towards the triplet ground state, a reasonable branching ratio allows a two infrared photon spontaneous cascading towards the singlet ground state *via* the 0_u^+ state. This Hund's "c" case state is a mixture of the $(1)^1\Sigma_u^+$ - $(1)^3\Pi_u$ states coupled by the spin-orbit interaction. This sample of deeply bound ground state molecules is an excellent starting point to perform laser cooling of the other degrees of freedom (vibrational and rotational), as will be discussed in the next sections.

3. TRAPPING

The development of electromagnetic techniques to trap molecules at sub-mK temperature is a first rank goal of the cold atomic and molecular community. An efficient molecular trap would be a great tool to explore molecular interactions where chemistry is dominated by pure quantum phenomena such as resonances, tunneling, s-wave scattering or collective effects. Furthermore, trapping cold molecules might open exciting possibilities for new research in molecular physics e.g.for measuring radiative lifetimes or for studying scattering dynamics.

For ultra-cold cesium dimers, two kinds of traps have been demonstrated in our group: a magnetic one suitable to trap only paramagnetic dimers and an optical one suitable to trap cesium molecules in any vibrational level of both electronic ground states.

3.1. Magnetic Trapping of Triplet Ground State Molecules

In ref. [134] we have reported an experimental setup where cold molecules formed *via* photoassociation in the $a^3\Sigma_u^+$ ground state were accumulated and trapped. This has been realized using a mixed atomic and molecular trap, consisted of a Cs vapor cell magneto-optical trap and a quadrupole magnetic Cs_2 trap, using the same magnetic field gradient. We have observed the trapping of 2.10^5 molecules formed and accumulated in the metastable $a^3\Sigma_u^+$ state at a temperature of 30±10 μK. As also shown in Fig. **9**, the accumulation saturates at a ≈ 150 ms photoassociation phase. The trapped molecules are distributed over many vibrational and some low rotational levels. The accumulation is limited by an optical pumping process; the PA light which forms the trapped molecules might also after 150 ms excite them toward excited states which can decay in either untrapped states or simply in states that are not detected. Finally, if all laser sources (PA or MOT laser) are turned off, the lifetime of the trapped molecular cloud appears to be limited by collisions with the Cs background gas to a value of the order of one second.

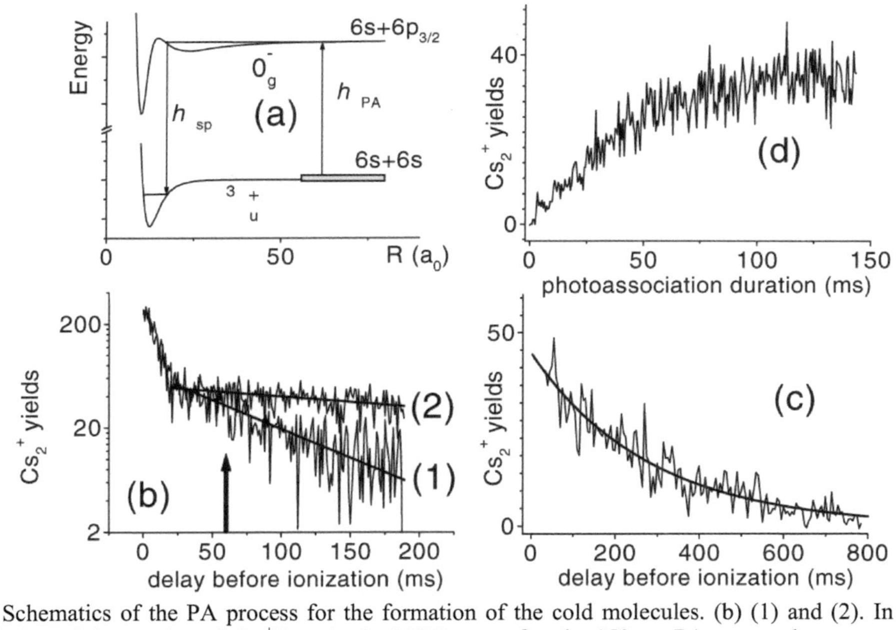

Figure 9: (a) Schematics of the PA process for the formation of the cold molecules. (b) (1) and (2). In vertical log scale, the evolution of the number of Cs_2^+ ions in the MOT zone after the 150 ms PA process has stopped at t=0: (1) MOT laser beams switched on. (2) MOT laser beams switched off; (c) Same as (b2) but with a higher background gas pressure and longer time observation of the trapping, at 0.5 Hz repetition rate and (solid line) exponential decay fitted life time. (d) The accumulation process observed 60 ms after the end of the PA process [see arrow in (b)]. This Figure is extracted from ref. [134].

3.2. Trapping in a Quasi-Electrostatic Dipole Trap

The main limitation of the magnetic trap is the fact that it works only for triplet molecules. The $X^1\Sigma_g^+$ ground electronic state of Cs_2 is not sensitive to magnetic field and can not be trapped using the same setup.

This is the main reason why we have also developed in Ref. [135] a trapping scheme able to trap all kinds of molecular states namely a dipole trap [136] provided by a focalized. With this tool it was possible to investigate cold inelastic collisions between confined cesium atoms and Cs_2 molecules.

Inelastic atom-molecule collisions have been observed and measured with a rate coefficient of ~ 2.5 x 10^{-11} cm^3 s^{-1}, mainly independent of the molecular rovibrational state populated. Here again the lifetimes of purely atomic and molecular samples were essentially limited by collisions with the background gas. The pure molecular trap lifetime ranges between 0,3-1 s, four times smaller than the atomic trap lifetime (see Fig. **10**) for the same conditions. This lower trapping time for the molecular species with respect to atoms was also observed in a pure magnetic trap. This fact has been interpreted as a signature of atom-molecule collisions, leading to destruction by dissociation of the molecule. In Fig. **10** and in ref. [135] we estimate the inelastic molecule-molecule collision rate to be ~ $10^{-11}cm^3s^{-1}$.

Dipole traps for ultracold cesium dimers have so far been realized also in two other groups. Their results can be found in Refs. [137, 138]. In these two trapping experiments the molecules were formed using the photoassociation schemes through the 0_g^- ($6s + 6p_{3/2}$) state, so the molecules were in the $a^3\Sigma_u^+$ electronic state and in relatively high vibrational levels. An interesting challenge was obviously to prepare the molecules in low vibrational levels of the singlet state which may exhibit different behavior against collisional processes. This was our goal that we have reached recently using vibrational cooling techniques.

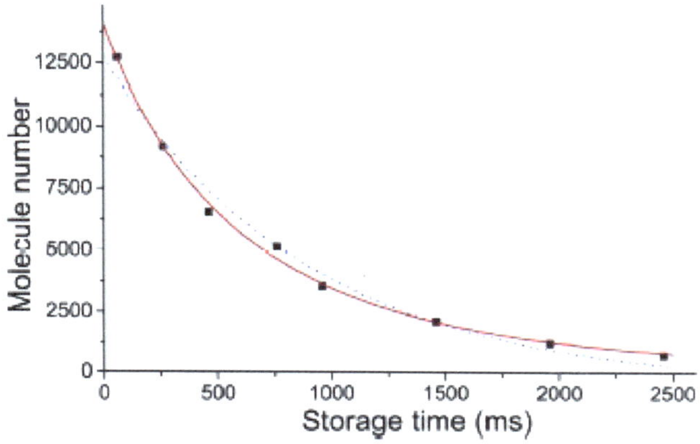

Figure 10: Trap lifetime of a pure molecular sample. The evolution of the number of molecules formed *via* PA of the 0_g^- ($6s+6p_{3/2}$) (v=6, J=2) state, is plotted (filled squares) and fitted with a function accounting for two-body collisions (solid line). A fit with an exponential decay function is also represented (dotted line). The fit yields a lifetime of 900 ms and a two-body rate coefficient of G_{Cs_2} =1.0(5).10^{-11} cm^3s^{-1}. This Figure is extracted from Ref. [135].

4. VIBRATIONAL COOLING BY OPTICAL PUMPING

Different schemes have been proposed to favor the formation of cold molecules in their lowest vibrational level. A few v = 0 (no vibration) ultracold ground-state potassium dimers have been observed [139] by two-photon PA, but several other vibrational levels were populated as well. By transferring a given vibrational level to the lowest vibrational one, cold ground state RbCs molecules have been prepared [140]. Recently LiCs molecules in the fundamental ground state have been observed after photoassociation [46]. Raman PA for preparing ultracold molecules in a well defined level has been studied by different groups. Its effiency is unfortunately limited, because the so-prepared molecules can be excited again and spontaneously decay toward other vibrational levels. Very recently, ultracold cesium molecules in the rovibronic v = 0 level (neither vibration nor rotation) have been formed using two Stimulated Raman Adiabatic Passages (STIRAP) on an ultracold atomic sample. The molecules are created in an optical lattice to prevent collision, and trapped for nearly 8 s [141].

Several theoretical approaches have been proposed as well to favor the spontaneous emission towards the lowest rovibrational level. For instance, the use of an external cavity has been proposed in ref.[142]. The vibration of the molecule can also be manipulated through quantum interferences between the different transitions. Interplay of control laser fields and spontaneous emission has been investigated for rotational or vibrational cooling [143-145].

Our approach towards $v_X = 0$ level of the ground state somehow follows the latter theoretical proposition of using a shaped laser. However, in our case the coherence of the field does not play any role since it is a simple incoherent optical pumping process that uses femtosecond pulses spectrally broad enough to excite all relevant vibrational levels.

4.1. Vibrational Cooling into the v_X=0 level by Simply Shaped Femtosecond Laser

In Section 2.4 we discussed the photoassociation scheme that allowed the efficient formation of deeply bound ultracold molecules distributed on the $v_X = 1 - 12$ vibrational levels of the singlet X $^1\Sigma_g^+$ ground state. In the following we demonstrated the transfer of this population towards the level $v_X = 0$, with no vibration [146]. The main idea is to use a broadband femtosecond laser tuned to the transitions $X^1\Sigma_g^+$ (v_X) towards $B^1\Pi_u$ (v_B) between the different vibrational levels of the ground state and of the electronically excited one. With the successive laser pulses, the absorption - spontaneous emission cycles lead through optical pumping to a redistribution of the vibrational population in the ground state, as sketched in Fig. **11c**. By shaping the laser to remove frequencies corresponding to the excitation of the $v_X = 0$ level (Fig. **11a**), this state become dark and molecules are accumulated in the $X^1\Sigma_g^+$ ($v_X = 0$), meaning vibrational laser cooling (Fig. **11b**).

The fs laser is a Tsunami mode locked laser from Spectra Physics with repetition rate 80 MHz, average power 2 W, pulse duration 100 fs and σ-Gaussian bandwidth 54 cm^{-1}. By tuning the laser wavelength, low vibrational levels ($v_X < 10$) are excited to vibrational levels v_B of the B state that through optical pumping redistribute the populations in the different vibrational levels of the ground state. The transitions from the v_X =0 level towards v_B levels are at frequencies superior to 13030 cm^{-1}.

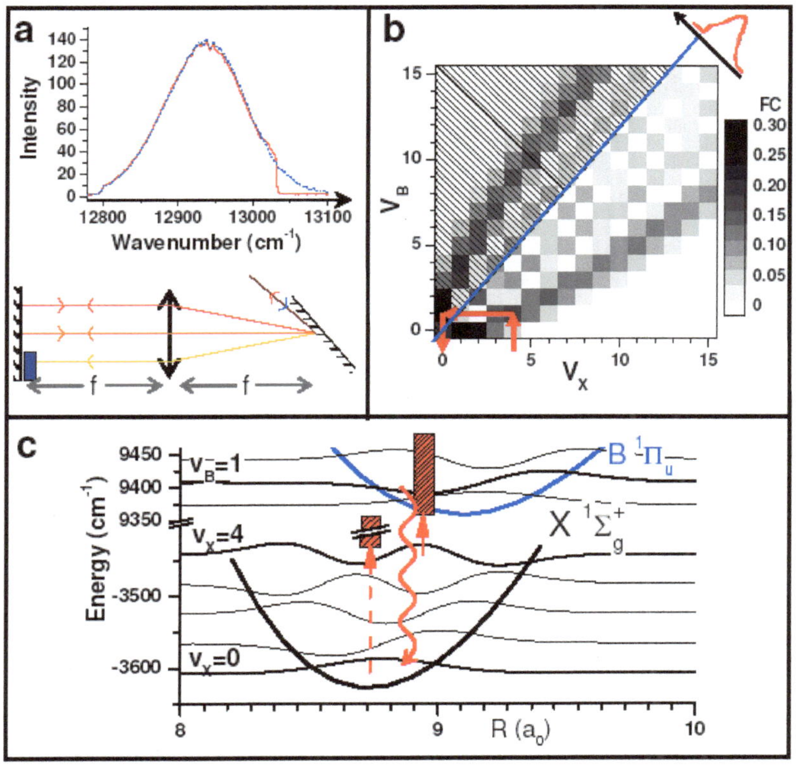

Figure 11: (a) Shaping of the femtosecond laser (see text). (b) Franck-Condon parabola indicating the importance of the Franck-Condon factor (level of grey). The shaded area, limited by the blue line, corresponds to missing laser frequencies due to shaping. (c) Scheme of the optical pumping. This Figure is extracted from ref. [127].

The spectral shape of the laser does not contain any transition from $v_X = 0$ (see Fig. **11b**). Our home-made shaper is a simple 4-f line using a grating (1800 lines per mm) for diffracting the laser beam. In this way, the vibrational level $v_X = 0$ becomes a dark state for the so-shaped laser. Fig. **11b** shows the Condon parabola of the Franck-Condon factors for the X $^1\Sigma_g^+$ (v) → B $^1\Pi_u$ (v') transitions. If we consider, for instance, $v_X = 4$, it is essentially excited in $v_c = 1$, which decays with a rate of about 30% to the dark level, $v_X = 0$, and with a rate of 70% essentially to the levels $v_X = 3$, 4, or 5. More generally, after a few cycles of absorption of laser light followed by spontaneous emission, a large fraction of the molecules can be accumulated in the lowest vibrational level $v_X = 0$.

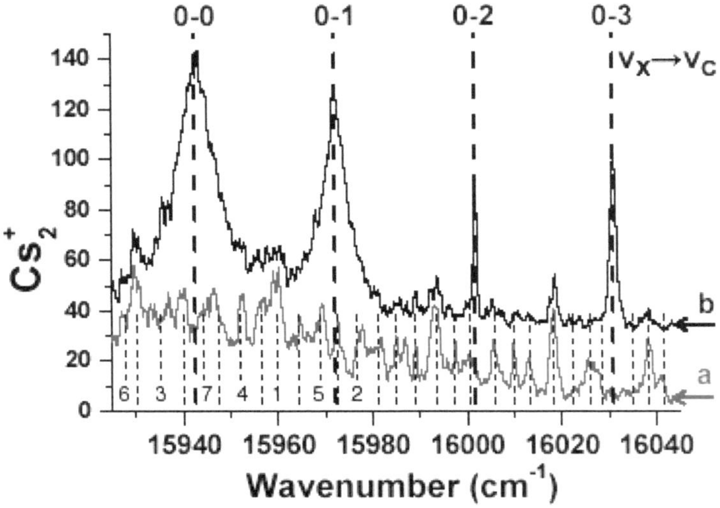

Figure 12: Cs_2^+ ion spectra: (a) without the shaped laser, (b) with the shaped laser. The numbers indicate vibrational levels v_X. The dashed lines indicate the resonance lines for vibrational transitions (v_X towards v_C) between the X $^1\Sigma_g^+$ and the C $^1\Pi_u$ states. This Figure is extracted from ref. [127].

Fig. **12b** shows the experimental results. The shaped laser strongly modifies the spectrum. The resonance lines corresponding to the transition $v_X = 0$ —> $v_C = 0$ — 3, mostly absent in the spectrum of Fig. **12a**, are in Fig. **12b** very strong. Their broadening corresponds to the saturation of the resonance in the REMPI process. The intensity of the lines indicates a very efficient transfer of the molecules in the lowest vibrational level, meaning a vibrational laser-cooling of the molecules. Fig. **13a** shows the experimental time evolution of the population in the different vibrational levels. Population transfer in the $v_X = 0$ level is almost saturated after the application of 1000 pulses, which requires ten microseconds. Taking into account the efficiency of the detection (< 10%), the detected ion signal corresponds to about one thousand molecules in the $v_X = 0$ level, corresponding to a formation rate of $v_X = 0$ molecules of more than 10^5 per second.

4.2. Theoretical Model of vibrational Cooling

We have modified the optical pumping in a very simple way. Using the known X $^1\Sigma_g^+$ and B $^1\Pi_u$ potential curves and their rotational constants [147,148], we have calculated the rovibrational energy levels. In the perturbative regime, we assume that the excitation probability is simply proportional to the laser spectral density at the transition frequencies, to the Franck-Condon factor [148], and to the Hönl-London factor [149]. We assume a laser spectrum shape very close to the experimental one: average intensity of 150 mW/cm^2 Gaussian shape center at 12940 cm^{-1} with a Gaussian line width σ = 54 cm^{-1}, and we removed all spectral components, due to the shaping above 13030 cm^{-1} After being excited by a pulse and before the shot of the next pulse, we assume a total decay of the excited state population with branching ratio given by the Franck-Condon factors, and the Hönl-London factors. The perturbative regime and the lifetime of the electronically excited state ~ 15 ns, close to the period of the pulses 12.5 ns, make reasonable the hypothesis of neglecting any accumulation of coherence due to the excitation by a train of ultrashort pulses [150]. This simple model shows that the molecules make a random walk, mostly in low vibrational levels, until reaching the $v_X = 0$ vibrational level. The accumulation of many molecules in the lowest vibrational level occurs with nearly unit transfer efficiency. Fig. **13b** shows a simulation of the transfer of the 70%

population into the $v_X = 0$ level, after 1000 pulses when the molecules are initially in a distribution of vibrational levels simulating the experimental ones. The model agrees well with the experimental data. It indicates that only about 5 absorption-spontaneous emission cycles (the number of necessary laser pulses depends on its intensity) are enough to transfer into $v_X = 0$ all molecules initially in $v_X < 10$ vibrational levels. The simulation shows also that the limitation of this mechanism is the optical pumping towards high vibrational levels. A broader bandwidth laser could probably increase the population in $v_X = 0$. Nevertheless, the Franck-Condon factors favor the accumulation of population in low vibrational levels.

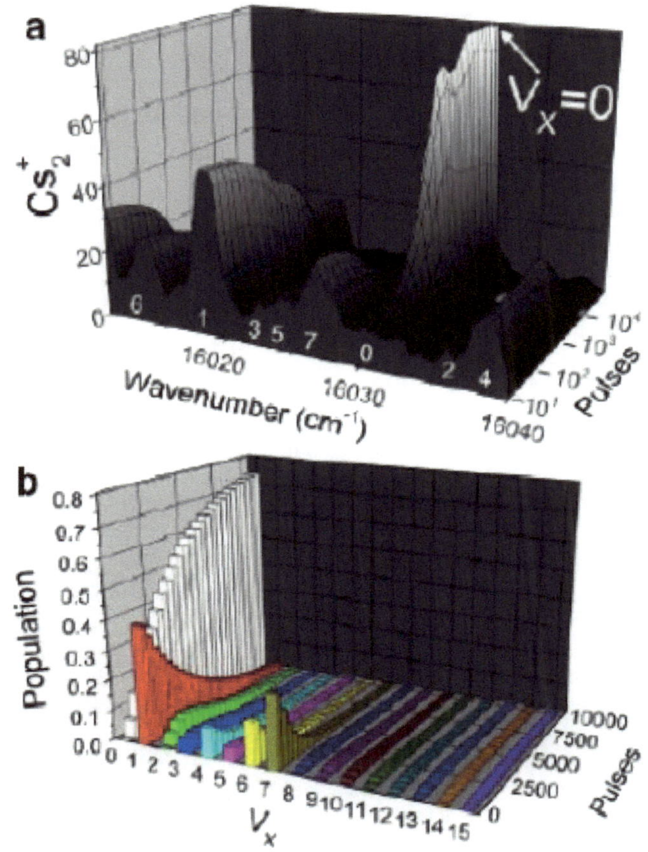

Figure 13: (a) Temporal evolution of the population transfer. (b) Simulation of the vibrational laser cooling. This Figure is extracted from ref. [127].

4.3. Vibrational Cooling into a Target v_X = 1-7 Level

The simple spectral cut used to optically pump population towards $v_X = 0$ can be generated to accumulate molecules into an arbitrary selected "target" vibrational level. However, in this case the spectral cut has to be more precise in the Fourier plane of the 4-f line because the frequencies needed to be removed are inside the spectrum and not only on one side of it. This can be done, for instance, by using a suitable selective element in the Fourier plane of the 4-f line because the frequencies needed to be removed are inside the spectrum and not only on one side of it. This can be done, for instance, by using a suitable selective element in the Fourier plane. We have first calculated, on the basis of known spectroscopy, the wavelengths to be removed from the laser spectrum and then implemented this by using ultrashort pulse shaping techniques based on a liquid crystal spatial light modulator (LC-SLM) as shown in Fig. **14** and reported in ref. [151]. A double LC-SLM is placed in the focal plane of the 4-f line, providing the correct amplitude shaping of the laser pulse before it is sent onto the molecular sample. As usually, the cold molecular cloud is created by PA using the scheme illustrated in Section 2.4 and then detected using the selective REMPI scheme, as in Fig. **12**.

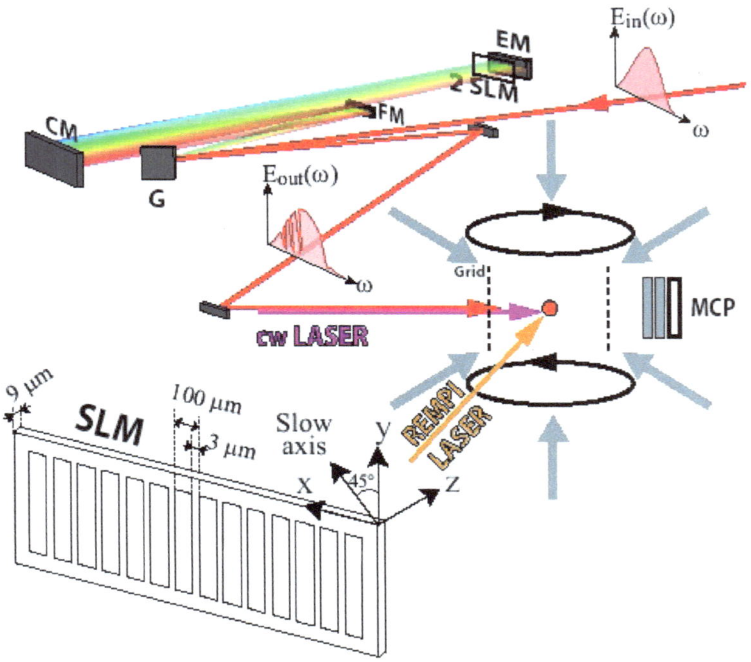

Figure 14: Experimental setup for the pulse shaper and the cold molecule production and detection. Upper part: Folded zero-dispersive line. The beam is dispersed by the grating G and then each spectral component is spatially separated and focused by the cylindrical mirror CM at the Fourier plane. FM is a plane folding mirror. Both LC-SLMs (detailed in the lower left side of the figure) are at the Fourier plane. An end mirror EM is placed just after the LC-SLM and the beam goes twice through the half of the line. The Figure is extracted from ref. [152].

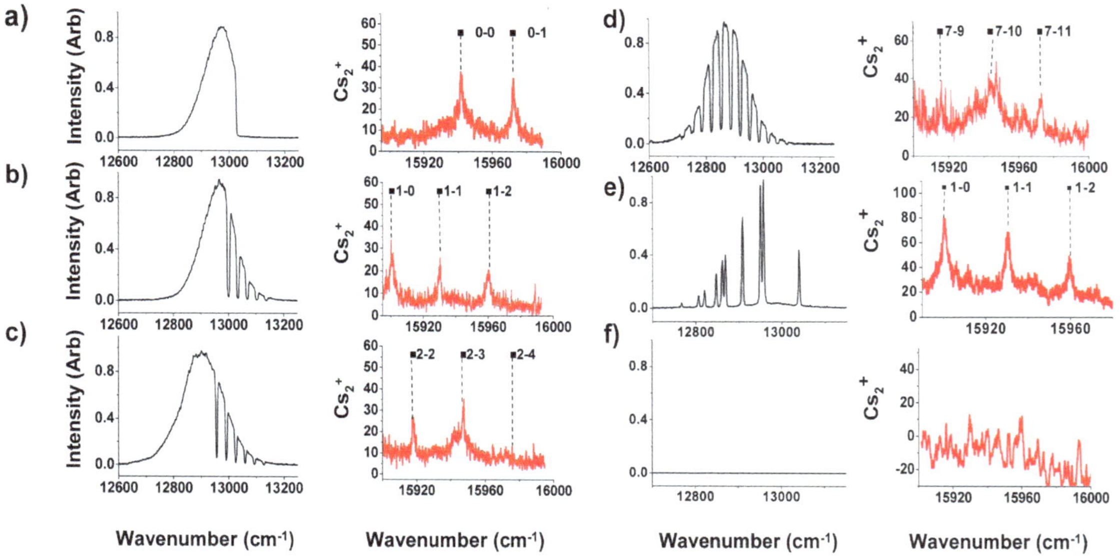

Figure 15: Experimental pulse spectra (black lines) used to cool the vibrational distribution into selected target levels and corresponding ionization spectra (red lines). Cases correspond to $v_X = 0$ (a), $v_X = 1$ (b), $v_X = 2$ (c), $v_X = 7$ (d) and $v_X = 1$ (e) levels respectively. The experimental ionization spectra (red lines) show mainly transitions involving only the target levels. Notice that in the spectrum (c) a small signal corresponding to molecules in the $v_x = 0$ level remains, and that in the spectrum (d) molecules in the $v_B = 1$ level also remain. In (e) the excitation spectrum contains only a few frequencies, calculated in order to pump molecules in an optimal way into $v_X = 1$. In (f), where the femtosecond laser is not present, all observed peaks can be assigned to specific $v_X \rightarrow v_B$ transitions. This Figure is extracted from ref. [152].

By removing the laser frequencies corresponding to the excitation of a selected v_X level, we make it impossible to pump molecules out of this level, thus making v_X a dark state. As time progresses a series of absorption – spontaneous emission cycles leads to accumulation of the molecules in this v_X level. We have demonstrated the technique for v_X = 0,1,2 and 7. In each case, as shown in Fig. **15**, a large fraction of the initially present molecules is transferred into a selected vibrational level.

Another approach to the shaping that we have demonstrated and that is illustrated in Fig. **15e** consisted in removing the most of the laser spectrum while maintaining only those frequencies that would excite any level (except the dark one) to the corresponding excited one with maximum probability of decaying into the target level. This kind of shaping has been called "optimum" shaping as it minimizes the number of absorption-spontaneous emission cycles necessary to pump the molecules into the target level, although not necessary time. This "optimized" strategy should reveal important when simultaneously trying to laser cool the rotational degree of freedom of molecules.

4.4. Vibrational Cooling with an Incoherent Diode Laser

One of important drawbacks of the system presented in the previous Section that could limit somehow its generalization is its complexity and cost, as it involves two expensive systems: a femtosecond laser and a LC-SLM. Another limitation is the non-perfect on/off ratio due to the gap between pixels in the LC-SLM. We demonstrated that both problems can be eliminated by using a simple broadband diode laser for the light source and a simple physical mask for the shaper. Different fixed masks can be easily produced by microfabrication but we are also currently working on using a micro-mirror's type of SLM. However, in a preliminary experiment [153], we have simply demonstrated the pumping process by using 4 needles placed by hands in the Fourier plane of the 4-f shaper line in order to suppress the main undesired frequencies and pump molecules into the v_X = 1 target level. We have also, with the broadband diode laser, easily recovered the preceding result of vibrational pumping into the v_X = 0 level. For the other target levels a more refined mask is required.

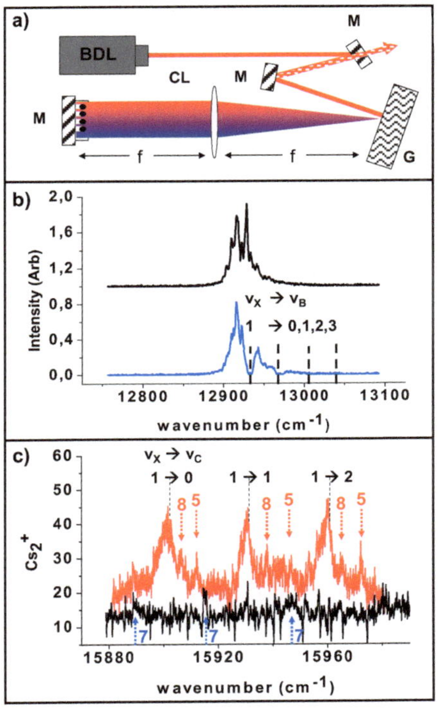

Figure 16: (a) The broadband diode laser shaping realized with the folded 4-f line setup consisted of three mirrors (M), a grating (G), a cylindrical lens (CL) and a mask. (b) Spectrum of the diode source before (upper line) and after (lower line) passing the 4-f line. The frequencies of v_X=1 $\rightarrow v_B$ =0, 1, 2, 3 transitions are removed. (c) The REMPI spectrum of the X $^1\Sigma_g^+ \rightarrow$ C $^1\Pi_u$ transitions with (upper trace, offset of 10 ions, for clarity) and without (lower trace) optical pumping with label to v_X levels and to $v_X \rightarrow v_C$ transitions. This figure is extracted from ref. [153].

As shown in Fig. **16**, we have demonstrated selective vibrational population transfer in cold cesium dimers using this simple approach based on the use of a shaped incoherent broadband diode laser. The broadband spectrum of the laser is wide enough to electronically excite several vibrational states of the molecule simultaneously and the pumping efficiency we obtained is similar to that of the femtosecond laser case.

An interesting outcome of this experiment is the fact that cooling towards $v_X = 1$ could be performed also with the diode laser operating below threshold, showing that coherence related effects are not relevant to this process. In this case the diode laser emits a luminescent pattern, like a collimated Light Emitting Diode (LED).

As the coherence of the light is not concerned, other light sources than a diode laser could be used. The requirements to be met are the large spectral bandwidth and the possibility to collimate the source enough to shape and focus it on the molecular sample.

5. PROSPECTS FOR ROTATIONAL COOLING

The use of optical pumping to cool molecular rotation has been discussed in some previous publications of our group [151, 152, 154, 155]. An important aspect of the molecular cooling *via* optical pumping is the effect of the vibrational cooling laser field on the rotation and *vice versa* [154]. The cooling fields discussed in this chapter treat the vibration and the rotation separately, *i.e.* when vibrational cooling is considered, the rotational degrees of freedom are omitted. As an example, in Fig. **17a** a simulation of the vibrational

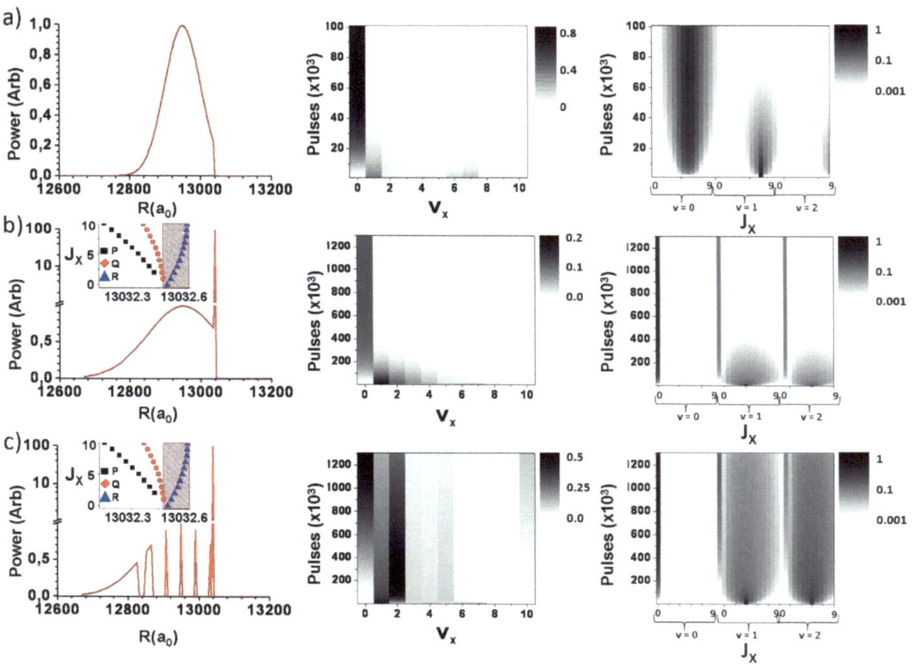

Figure 17: (color online) (a) A simulation of the vibrational cooling process with the use of a simply shaped pulse shown on the left. In the middle, we see the modification in the vibrational distribution and on the right the result on the rotational distribution, (b) Attempt to simultaneously cool the vibration and the rotation with a simply shaped pulse and an additional rotational cooling field (left). The high intensity peak is the field designed to cool rotation exciting only the P and Q branches (see inlet). The evolution of the vibrational distribution (middle) and the rotational distribution (right) shows that the simply shaped pulse cannot efficiently compensate the vibrational heating effect, (c) The situation is improved if optimized pulses are considered for the vibrational cooling, and almost 50% of the molecular population is now accumulated in the absolute ground state $v_X = J_X = 0$. The Figure is extracted from ref. [152].

cooling in cesium dimers with the use of a simply shaped pulse is shown in the left part. In the middle, the modification induced on the vibrational distribution is shown, while in the right part we show the rotational distribution in the first three vibrational levels. In this simulation, we consider ten rotational levels in each vibrational state, while the molecular distribution is initially in the state with J =5. The rotational distribution is spreading under the influence of the vibrational cooling field. The only reason that this effect is not reflected in the experimental data, is that the pulsed REMPI detection is broadband enough in order to simultaneously probe all rotational levels. A similar optical pumping process designed to cool the molecular rotation would have a similar heating effect on the vibration. However, the molecular population would be spread in much larger energy and its simultaneous detection would be impossible. Thus, a field designed to cool the molecular rotation *via* optical pumping should be accompanied by a vibrational cooling field sufficient to cancel the vibrational heating effect.

A natural possibility for laser cooling of the molecular rotation is to consider a laser field shaped so that the frequencies corresponding to the transitions $\Delta J = J_e — J_X = 0, +1$ (Q and R bands) are removed (here J_e and J_X represent the rotational quantum numbers in the excited and ground levels respectively), and leave only the $\Delta J = J_e — J_X = -1$ (P band) ones that are J lowering (except the $J_X = 1 \rightarrow J_e = 1$ that is needed to empty the $J_X = 1$ level). With such a shaping, absorption- spontaneous emission cycles would indeed lead to a decrease on average of the principal rotation quantum number J_X, *i.e.* to laser cooling of rotational degree of freedom.

However in Fig. **17b** we see that a simply shaped pulse is not enough to efficiently compensate the heating effect of the rotational cooling field since less than 20% of the molecular population is finally transferred to the ground rovibrational state. On the contrary, if an optimized pulse of the same bandwidth is considered for the vibrational cooling, the situation is highly improved as ~50% of the molecular population is accumulated in the ground rovibrational level, as shown in the simulation displayed in Fig. **17c**.

The efficiency of such "optimized" vibrational cooling scheme has been experimentally demonstrated in the case of $v_X = 1$ as shown in Fig. **15e**, where the excitation spectrum contains only a few frequencies, calculated in order to pump molecules in an optimal way (i,e, using the fewer number of photon absorptions).

5.1. Rotational Analysis by Depletion Spectroscopy

If the population of a specific ground-state vibrational level can be monitored by one-color two- photon pulsed-laser ionization, this is not the case for the rotational population because the laser linewidth is larger than the rotational spacing. Furthermore the amount of laser power used for ionization creates a large power broadening that would nevertheless hide the rotational sub-structure. One solution to resolve and monitor the rotational structure is the so called depletion spectroscopy (see Fig. **18**).

When a cw laser is resonant with a rovibrational transition to an excited state, the ground state population, and hence the ion signal, is depleted. This narrowband spectroscopic technique allows the individual rotational levels in both ground and excited states to be resolved, and thus the population of a single ground state rovibrational level to be monitored.

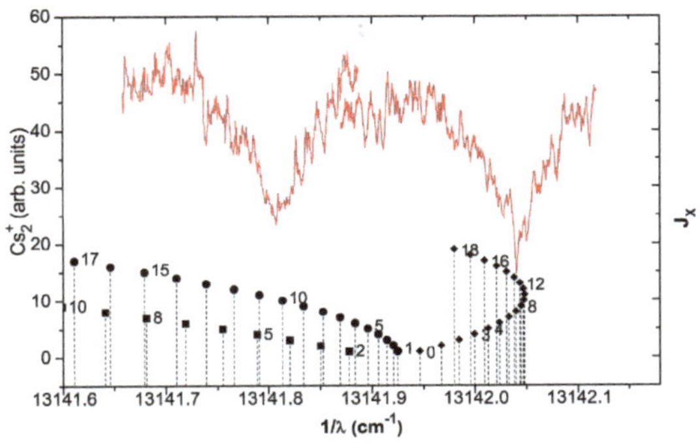

Figure 18: In red, depletion spectra of the ground state $v_X = 0$ level. The depletion laser is scanned on $v_X = 0 \rightarrow v_B = 3$ transition. Two different scans are necessary to cover the whole frequency range. In blank, the calculated transitions (shown as vertical lines with associated J_X values). For the P (squares), Q (circles) and R (diamond) branches are based on experimental data [147, 148] This figure is extracted from ref. [153].

6. CONCLUSIONS

In this chapter we have briefly stated the main motivations for the study and the realization of cold (T < 1 mK) molecular samples. We have given a comprehensive list of all cold molecules formed up to now with the relevant references. Among the methods for producing cold molecules we have illustrated the photoassociation of cold atoms. For the cesium case we have then reviewed the formation, cooling and trapping of the molecules either in highly or in deeply bound singlet or triplet state. We have also detailed the detection procedure.

The extension of this method to cool the rotational degrees of freedom is possible as long as the required resolution to the shaping can be achieved. Its generalization to molecular species other than cesium dimers is surely possible but requires careful choice of molecular states (study of molecular potential curves) and the availability of broadband collimated light sources in the required energy range. Preliminary studies and ongoing experiments concern by now the NaCs cold dimer.

The vibrational cooling process does not depend on the temperature as long as the molecule has the time to interact with the pumping light and no quenching processes takes place. So, additional and more important advances that can be foreseen for such a method is the use of an antireflection coated broad area laser diode. Such optical pumping can be used as a repumper light in a scheme for a direct laser cooling of molecules. The relatively low cost and the simplicity of such a setup allows for the use of more than one diode laser if needed, for instance, each of them addressing transitions between different vibrational levels or even different electronic states.

Similar results could also be reached for heteronuclear systems and the formation of polar molecules, opening exciting perspectives in quantum information.

Particularly interesting is the cooling of internal or external degrees of freedom of polar molecules loaded in an electrostatic trap, after velocity filtering of an effusive molecular beam [89, 156].

ACKNOWLEDGEMENTS

This chapter reports on the work of many PhD students, post-docs and.permanent or visiting researchers that spent some time in Laboratoire Aimé Cotton in the last ten years, Credit to any of them is given by citing the principal reference to each result. This work is supported in Orsay by the "Institut Francilien Recherche les Atomes Froids" (IFRAF). A.F. has been supported by the "*Triangle de la Physique*" under contracts 2007-n.74T and 2009-035T "GULFSTREAM".

REFERENCES

[1] Rabitz H, de Vivie-Riedle R, Motzkus M, Kompa K. Whither the future of controlling quantum phenomena? Science 2000; 288(5467): 824-28.

[2] Osborne I, Coontz R. Quantum wonderland. Science 2008; 319(5867): 1201.

[3] Dulieu O, Raoult M, Tiemann E. Cold molecules: a chemistry kitchen for physicists? J Phys B: At Mol Opt Phys 2006; 39(19): Introductory Review.

[4] Smith IWM, Ed. Low temperatures and cold molecules. London: Imperial College Press 2008.

[5] Bell MT, Softley TP. Ultracold molecules and ultracold chemistry. Molecular Physics 2009; 107(2): 99-132.

[6] Carr LD, DeMille, D, Krems RV, Ye J. Cold and ultracold molecules: science, technology and applications. New J Phys 2009; 11(5): 055049.

[7] Dulieu O, Gabbanini C. The formation and interactions of cold and ultracold molecules: new challenges for interdisciplinary physics. Rep Prog Phys 2009; 72(8): 086401.

[8] Stwalley WC, Krems RV, Friedrich B. Eds. Cold molecules: theory, experiment, applications. Boca Raton, Florida: CRC Press 2009

[9] Krems RV. Cold controlled chemistry. Phys Chem Chem Phys 2008; 10(28): 4079-92.

[10] Schnell M, Meijer G. Cold molecules: preparation, applications, and challenges. Angew Chem Int Ed Engl 2009; 48(33): 6010.

[11] Chin C, Kraemer T, Mark M, *et al.* Observation of feshbach-like resonances in collisions between ultracold molecules. Phys Rev Lett 2005; 94(12): 123201.

[12] Takasu Y, Komori K, Honda K, *et al.* Photoassociation spectroscopy of laser-cooled ytterbium atoms. Phys Rev Lett 2004; 93(12): 123202.

[13] A non exhaustive list of the molecules is: Mg1-3HCN, Ne, Kr, ArHF, tetracene-Ar, HCN-H_2, HD, D_2, Ag8-Ne_N, Ar_N, Kr_N, Xe_N (N = 1-135), NaCs, LiCs, HF-(H_2), OCS-$(H_2)_N$ (N = 1-17), Cs_2, Rb_2, CO, HCl, amino acid, 3-hydroxyflavone, xanthine, $[Na(H_2O)_N]^+$ (N = 6-43) (Falconer TM, Lewis WK, Bemish RJ, Miller RE, Glish GL. Formation of cold ion-neutral clusters using superfluid helium nanodroplets. Rev Sci Instrum 2010; 81(5): 054101). HF-N_2O, Mg-HF, Mg-$(HF)_2$, CH_3-H_2O, Cs_N^+ $(H_2O)_M$, CF_3I, CH_3I.

[14] Even U, Jortner J, Noy D, Lavie N, Cossart-Magos C. Cooling of large molecules below 1 K and He clusters formation. J Chem Phys 2000; 112(16): 8068-71.

[15] Choi MY, Douberly GE, Falconer TM, *et al.* Infrared spectroscopy of helium nanodroplets: novel methods for physics and chemistry. Int Rev Phys Chem 2006; 25(1): 15-75.

[16] Stienkemeier F, Lehmann KK. Topical review: spectroscopy and dynamics in helium nanodroplets. J Phys B: At Mol Opt Phys 2006; 39(1): 127-44.

[17] Barranco M, Guardiola R, Hernandez S, *et al.* Helium nanodroplets: an overview. J Low Temp Phys 2006; 142(1-2): 1-81.

[18] Küpper J, Merritt JM. Spectroscopy of free radicals and radical containing entrance-channel complexes in superfluid helium nanodroplets. Int Rev Phys Chem 2007; 26(2): 249-87.

[19] Chin C, Grimm R, Julienne P, Tiesinga E. Feshbach resonances in ultracold gases. Rev Mod Phys 2010; 82(2): 1225-86.

[20] Köhler T, Goral K, Julienne PS. Production of cold molecules *via* magnetically tunable Feshbach resonances. Rev Mod Phys 2006; 78(14): 1311-61.

[21] Donley EA, Claussen NR, Thompson ST, Wieman CE. Atom-molecule coherence in a Bose-Einstein condensate. Nature (London) 2002; 417(6888): 529-33.

[22] Thompson ST, Hodby E, Wieman CE. Ultracold molecule production *via* a resonant oscillating magnetic field. Phys Rev Lett 2005; 95(19): 190404.

[23] Papp SB, Wieman CE. Observation of heteronuclear feshbach molecules from a ^{85}Rb ^{87}Rb gas. Phys Rev Lett 2006; 97(18): 180404.

[24] Herbig J, Kraemer T, Mark M, *et al.* Preparation of a pure molecular quantum gas. Science 2003; 301(5639): 1510-3.

[25] Greiner M, Regal CA, Jin DS. Emergence of a molecular Bose-Einstein condensate from a Fermi gas. Nature (London) 2003; 426(6966): 537-40.

[26] Jochim S, Bartenstein M, Altmeyer A, *et al.* Bose-Einstein condensation of molecules. Science 2003; 302(5653): 2101-4.

[27] Cubizolles J, Bourdel T, Kokkelmans SJ, Shlyapnikov GV, Salomon C. Production of long-lived ultracold Li_2 molecules from a Fermi gas. Phys Rev Lett 2003; 91(24): 240401.

[28] Strecker KE, Partridge GB, Hulet RG. Conversion of an atomic Fermi gas to a long-lived molecular bose gas. Phys Rev Lett 2003; 91(8): 080406.

[29] Xu K, Mukaiyama T, Abo-Shaeer JR, Chin JK, Miller DE, Ketterle W. Formation of quantum-degenerate sodium molecules. Phys Rev Lett 2003; 91(21): 210402.

[30] Ospelkaus C, Ospelkaus S, Humbert, L, Ernst P, Sengstock K, Bongs K. Ultracold heteronuclear molecules in a 3d optical lattice. Phys Rev Lett 2006; 97(12): 120402.

[31] Weber C, Barontini G, Catani J, Thalhammer G, Inguscio M, Minardi F. Association of ultracold double-species bosonic molecules. Phys Rev A 2008; 78(6); 061601.

[32] Beaufils Q, Crubellier A, Zanon T, *et al.* Feshbach resonance in d -wave collisions. Phys Rev A 2009; 79(3): 032706.

[33] Lompe T, Ottenstein TB, Serwane F, Wenz AN, Zurn G, Jochim S. Radio frequency association of Efimov trimers. ArXiv e-prints 2010.

[34] Jones KM, Tiesinga E, Lett PD, Julienne PS. Ultracold photoassociation spectroscopy: Long-range molecules and atomic scattering. Rev Mod Phys 2006; 78(2): 483-535.

[35] Fioretti A, Comparat D, Crubellier A, Dulieu O, Masnou-Seeuws F, Pillet P. Formation of cold Cs_2 molecules through photoassociation. Phys Rev Lett 1998; 80(20): 4402-5.

[36] Mosk AP, Reynolds MW, Hijmans TW, Walraven JTM. Photoassociation of spin-polarized hydrogen. Phys Rev Lett 1999; 82(2): 307-310.

[37] Gabbanini C, Fioretti A, Lucchesini A, Gozzini S, Mazzoni M. Cold rubidium molecules formed in a magneto-optical trap. Phys Rev Lett 2000; 84(13): 2814-17.

[38] Gerton JM, Strekalov D, Prodan I, Hulet RG. Direct observation of growth and collapse of a Bose-Einstein condensate with attractive interactions. Nature (London), 2000; 408(6813): 692-5.

[39] Fatemi FK, Jones KM, Lett PD, Tiesinga E. Ultracold ground-state molecule production in sodium. Phys Rev A 2002; 66(5): 053401.

[40] Nikolov AN, Enscher JR, Eyler EE, Wang H, Stwalley WC, Gould PL. Efficient production of ground state potassium molecules at sub-mK temperatures by two-step photoassociation. Phys Rev Lett 2000; 84(2): 246-9

[41] Gaebler JP, Stewart JT, Bohn JL, Jin, DS. p-Wave Feshbach molecules. Phys Rev Lett 2007; 98(20): 200403.

[42] Moal S, Portier M, Kim J, *et al.* Accurate determination of the scattering length of metastable helium atoms using dark resonances between atoms and exotic molecules. Phys Rev Lett 2006; 96(2): 023203.

[43] Zinner G, Binnewies T, Riehle F, Tiemann E. Photoassociation of cold Ca aAtoms. Phys Rev Lett 2000; 85(11): 2292-95.

[44] Wang D, Qi J, Stone MF, *et al.* Photoassociative production and trapping of ultracold KRb molecules. Phys Rev Lett 2004; 93(24): 243005.

[45] Kerman AJ, Sage JM, Sainis S, Bergeman T, Demille, D. Production and state- selective detection of ultracold RbCs molecules. Phys Rev Lett 2004; 92(15): 153001.

[46] Haimberger C, Kleinert J, Bhattacharya M, Bigelow NP. Formation and detection of ultracold ground-state polar molecules. Phys Rev A 2004; 70(2): 021402.

[47] Deiglmayr J, Grochola A, Repp M, *et al.* Formation of ultracold polar molecules in the rovibrational ground state. Phys Rev Lett 2008; 101(13): 133004.

[48] Zwierlein MW, Stan CA, Schunck CH, *et al.* Observation of Bose-Einstein condensation of molecules. Phys Rev Lett 2003; 91(25): 250401.

[49] Weinstein JD, Decarvalho R, Guillet T, Friedrich B, Doyle, JM. Magnetic trapping of calcium monohydride molecules at millikelvin temperatures. Nature (London), 1998; 395(5652): 148-50.

[50] Weinstein JD, Decarvalho R, Amar K, *et al.* Spectroscopy of buffer-gas cooled vanadium monoxide in a magnetic trapping field. J Chem Phys. 1998; 109(7): 2656-61.

[51] Maussang K, Egorov D, Helton JS, Nguyen SV, Doyle JM. Zeeman relaxation of CaF in low-temperature collisions with helium. Phys Rev Lett 2005; 94(12): 123002.

[52] Egorov D, Weinstein JD, Patterson D, Friedrich B, Doyle JM. Spectroscopy of laser-ablated buffer-gas-cooled PbO at 4 K and the prospects for measuring the electric dipole moment of the electron. Phys Rev A 2001, 63(3): 030501.

[53] Maxwell SE, Brahms N, Decarvalho R, *et al.* High-flux beam source for cold, slow atoms or molecules. Phys Rev Lett 2005; 95(17): 173201.

[54] Patterson D, Doyle JM. Bright guided molecular beam with hydrodynamic enhancement. J Chem Phys 2007; 126(15): 154307.

[55] Egorov D, Campbell WC, Friedrich B, *et al.* Buffer-gas cooling of NH *via* the beam loaded buffer-gas method. Eur Phys J D 2004; 31(2): 307-11.

[56] Stoll M, Bakker JM, Steimle TC, Meijer G, Peters A. Cryogenic buffer-gas loading and magnetic trapping of CrH and MnH molecules. Phys Rev A 2008, 78(3): 032707.

[57] Van Buuren LD, Sommer C, Motsch M, *et al.* Electrostatic extraction of cold molecules from a cryogenic reservoir. Phys Rev Lett 2009, 102(3): 033001.

[58] Skoff SM, Hendricks RJ, Sinclair CDJ, *et al.* Doppler-free laser spectroscopy of buffer-gas-cooled molecular radicals, New J Phys 2009, 11(12): 123026.

[59] Doyle J. Cold, trapped molecules *via* cryogenic buffer gas methods. APS Meeting Abstracts 2010, pp. 24001.

[60] Van de Meerakker SYT, Bethlem HL, Meijer, G. Taming molecular beams. Nature Physics, 2008; 4(8): 595–602

[61] Bethlem HL, Berden G, Meijer G. Decelerating Neutral Dipolar Molecules. Phys Rev Lett 1999; 83(8) 1558–61.

[62] Bethlem HL, Berden G, Crompvoets FMH, Jongma RT, Van Roij AJA, Meijer G. Electrostatic trapping of ammonia molecules. Nature (London) 2000; 406(6795): 491–4.

[63] Crompvoets FMH, Bethlem HL; Jongma RT, Meijer G. A prototype storage ring for neutral molecules. Nature (London) 2001; 411(6834): 174–6.

[64] Bochinski JR, Hudson ER, Lewandowski HJ, Meijer G, Ye J. Phase space manipulation of cold free radical OH molecules. Phys Rev Lett 2003; 91(24): 243001.

[65] Bochinski JR, Hudson ER, Lewandowski HJ, Ye J. Cold free-radical molecules in the laboratory frame. Phys Rev A 2004; 70(4): 043410.

[66] Van de Meerakker SY, Smeets PH, Vanhaecke N, Jongma RT, Meijer G. Deceleration and electrostatic trapping of OH radicals. Phys Rev Lett 2005; 94(2); 023004.

[67] Hudson ER, Ticknor C, Sawyer BC, *et al.* Production of cold formaldehyde molecules for study and control of chemical reaction dynamics with hydroxyl radicals. Phys Rev A 2006; 73(6): 063404.

[68] Van de Meerakker SYT, Labazan I, Hoekstra S, Kupper J, Meijer G. Production and deceleration of a pulsed beam of metastable NH (a $^1\Delta$) radicals. J Phys B: At Mol Opt Phys 2006; 39(5): S1077-84.

[69] Bucicov O, Nowak M, Jung S, *et al.* Cold SO_2 molecules by Stark deceleration. Eur Phys J D 2008; 46(3): 463-9.

[70] Wohlfart K, Gratz F, Filsinger F, Haak H, Meijer G, Kupper J. Alternating-gradient focusing and deceleration of large molecules. Phys Rev A 2008; 77(3): 031404.

[71] Tarbutt, MR, Bethlem HL, Hudson JJ, *et al.* Slowing heavy, ground-state molecules using an alternating gradient decelerator. Phys Rev Lett 2004; 92(17); 173002.

[72] Tokunaga SK, Dyne JM, Hinds EA, Tarbutt MR. Stark deceleration of lithium hydride molecules. N J Phys 2009; 11(5): 055038.

[73] Wall TE, Tokunaga SK, Hinds EA, Tarbutt MR. Nonadiabatic transitions in a Stark decelerator. Phys Rev A 2010; 81(3): 033414.

[74] Vliegen E, Worner HJ, Softley TP, Merkt F. Nonhydrogenic effects in the deceleration of Rydberg atoms in inhomogeneous electric fields. Phys Rev Lett 2004; 92(3): 033005.

[75] Yamakita Y, Takahashi R, Ohno K, Procter SR, Maguire G, Softley TP. Cooling effects in the Stark deceleration of Rydberg atoms/molecules with time-dependent electric fields. J Phys Conf Ser 2007; 80: 012045.

[76] Hogan SD, Merkt F. Demonstration of three-dimensional electrostatic trapping of state-selected Rydberg atoms. Phys Rev Lett 2008; 100(4): 043001.

[77] Hogan SD, SeilerC, Merkt F. Rydberg-state-enabled deceleration and trapping of cold molecules. Phys Rev Lett 2009; 103(12): 123001.

[78] Fulton R, Bishop AI Barker PF. Optical Stark decelerator for molecules. Phys Rev Lett 2004; 93 (24): 243004.

[79] Fulton R, Bishop AI, Shneider MN, Barker PF. Optical Stark deceleration of nitric oxide and benzene molecules using optical lattices. J Phys B: At Mol Phys 2006; 39(5): S1097-1109.

[80] Narevicius E, Libson A, Parthey CG, *et al.* Stopping supersonic oxygen with a series of pulsed electromagnetic coils: A molecular coilgun. Phys Rev A 2008; 77(5): 051401.

[81] Elioff MS, Valentini JJ, Chandler DW. SubKelvin cooling of molecules *via* "billiard- like" collisions with argon. Science 2003; 302(5652): 1940-3.

[82] Liu, NN, Loesch H. Kinematic slowing of molecules formed by reactive collisions. Phys Rev Lett 2007; 98(10): 103002.

[83] Kay JJ, Van de Meerakker SYT, Strecker KE, Chandler DW. Production of cold ND_3 by kinematic cooling. cold and ultracold molecules. Faraday Discussions 2009; 142: 143-54.

[84] Trottier A, Carty D, Wrede E. Photostop: production of zero-velocity molecules by photodissociation in a molecular beam. ArXiv e-prints. 2010.

[85] Gupta M, Herschbach D. Slowing and speeding molecular beams by means of a rapidly rotating source. J Phys Chem A 2001; 105(9): 1626.

[86] Gupta M, Herschbach D. A mechanical means to produce intense beams of slow molecules. J Phys Chem A 1999; 103(50): 10670-3.

[87] Strebel M, Stienkemeier F, Mudrich M. Improved setup for producing slow beams of cold molecules using a rotating nozzle. Phys Rev A 2010; 81(3): 033409.

[88] Deachapunya S, Fagan PJ, Major AG, *et al.* Slow beams of massive molecules. Eur Phys J D 2008; 46: 307-13.

[89] Rangwala SA, Junglen T, Rieger T, Pinkse PW, Rempe G. Continuous source of translationally cold dipolar molecules. Phys Rev A 2003; 67(4): 043406.

[90] Rieger T, Junglen T, Rangwala SA, *et al.* Water vapor at a translational temperature of 1K. Phys Rev A 2006; 73(6): 061402.

[91] Willitsch S, Bell MT, Gingell AD, Procter SR, Softley TP. Cold reactive collisions between laser-cooled ions and velocity-selected neutral molecules. Phys Rev Lett 2008; 100(4): 043203.

[92] Liu Y, Yun M, Xia Y, Deng L, Yin J. Experimental generation of a cw cold CH_3CN molecular beam by a low-pass energy filtering. Physical Chemistry Chemical Physics (Incorporating Faraday Transactions) 2010; 12(3): 745-52.

[93] Motsch M, Van Buuren LD, Sommer C, Zeppenfeld M, Rempe G, Pinkse PWH. Cold guided beams of water isotopologs. Phys Rev A 2009; 79(1): 013405.

[94] Tsuji H, Sekiguchi T, Mori T, Momose T, Kanamori H. Stark velocity filter for nonlinear polar molecules. J Phys B: At Mol Phys 2010; 43(9); 095202.

[95] Wester R. TUTORIAL: Radiofrequency multipole traps: tools for spectroscopy and dynamics of cold molecular ions. J Phys B: At Mol Phys 2009; 42(15); 154001.

[96] Wineland DJ, Monroe C, Itano WM, Leibfried D, King BE, Meekhof DM. Experimental issues in coherent quantum-state manipulation of trapped atomic ions. J Res Nat Inst Stand Tech 1998; 103(3): 259-328.

[97] Ostendorf A, Zhang CB, Wilson MA, Offenberg D, Roth B, Schiller S. Sympathetic cooling of complex molecular ions to millikelvin temperatures. Phys Rev Lett 2006; 97(24): 243005.

[98] Mølhave K, Drewsen M. Formation of translationally cold MgH^+ and MgD^+ molecules in an ion trap. Phys Rev A 2000; 62(1): 011401.

[99] Hornekær L. Single- and multi-species coulomb ion crystals: structures, dynamics and sympathetic cooling. PhD Thesis 2000, Aarhus University, The Netherlands.

[100] Blythe P, Roth B, Frohlich U, Wenz H, Schiller S. Production of ultracold trapped molecular hydrogen ions. Phys Rev Lett 2005; 95(18): 183002.

[101] Roth B, Offenberg D, Zhang CB, Schiller S. Chemical reactions between cold trapped Ba^+ ions and neutral molecules in the gas phase. Phys Rev A 2008; 78(4): 042709.

[102] Offenberg D, Wellers C, Zhang CB, Roth B, Schiller S. Measurement of small photodestruction rates of cold, charged biomolecules in an ion trap. J Phys B: At Mol Opt Phys 2009; 42(3):035101.

[103] Offenberg D, Zhang CB, Wellers C, Roth B, Schiller S. Translational cooling and storage of protonated proteins in an ion trap at subkelvin temperatures. Phys Rev A, 2008; 78(6): 061401.

[104] Offenberg D, Wellers C, Zhang CB, Roth B, Schiller S. Measurement of small photodestruction rates of cold, charged biomolecules in an ion trap. J Phys B: At Mol Opt Phys 2009; 42(3):035101.

[105] Lett PD, Helmerson K, Phillips WD, Ratliff LP, Rolston SL, Wagshul ME. Spectroscopy of Na_2 by photoassociation of laser-cooled Na. Phys Rev Lett 1993; 71(14): 2200-3.

[106] Miller JD, Cline RA, Heinzen DJ. Photoassociation spectrum of ultracold Rb atoms. Phys Rev Lett 1993; 71(14): 2204-7.

[107] Jones KM, Tiesinga E, Lett PD, Julienne PS. Ultracold photoassociation spectroscopy: Long-range molecules and atomic scattering. Rev Mod Phys 2006; 78(2): 483-535.

[108] Thorsheim HR, Weiner J, Julienne PS. Laser-induced photoassociation of ultracold sodium atoms. Phys Rev Lett 1987; 58(23): 2420-3.

[109] Lisdat C, Vanhaecke N, Comparat D, Pillet P. Line shape analysis of two-colour photoassociation spectra on the example of the Cs ground state. Eur Phys J D 2002; 21(3): 299-309.

[110] Vanhaecke N, Lisdat C, T'jampens B, Comparat D, Crubellier A, Pillet P. Accurate asymptotic ground state potential curves of Cs_2 from two-colour photoassociation. Eur Phys J D 2004; 28(2); 351-60.

[111] Mark M, Danzl J, Haller E, *et al.* Dark resonances for ground-state transfer of molecular quantum gases. Appl Phys B: Lasers O 2009; 95(2): 219-25.

[112] Danzl JG, Mark MJ, Haller E, *et al.* Precision molecular spectroscopy for ground state transfer of molecular quantum gases. cold and ultracold molecules. Faraday Discussions 2009; 142: 283-95.

[113] Dion CM, Drag C, Dulieu O, Laburthe Tolra B, Masnou-Seeuws F, Pillet P. Resonant coupling in the formation of ultracold ground state molecules *via* photoassociation. Phys Rev Lett 2001; 86(11): 2253-6.

[114] Pichler M, Chen H, Stwalley WC. Photoassociation spectroscopy of ultracold Cs below the $6p_{1/2}$ limit. J Chem Phys 2004; 121(4): 1796-1801.

[115] Fioretti A, Comparat D, Drag C, *et al.* Photoassociative spectroscopy of the Cs_2 0_g^- long-range state. Eur Phys J D 1999; 5(3): 389-403.

[116] Comparat D, Drag C, Fioretti A, Dulieu O, Pillet P. Photoassociative spectroscopy and formation of cold molecules in cold cesium vapor: trap-loss spectrum versus ion spectrum. J Mol Spectrosc 1999; 195(2): 229-35.

[117] Comparat D, Drag C, Laburthe Tolra B, *et al.* Formation of cold Cs_2 ground state molecules through photoassociation in the 1_u pure long-range state. Eur Phys J D 2000; 11(1): 59-71.

[118] Drag C, Laburthe Tolra B, T'jampens B, *et al.* Photoassociative spectroscopy as a self-sufficient tool for the determination of the Cs triplet scattering length. Phys Rev Lett 2000; 85(7): 1408-11.
[119] Drag C, Laburthe-Tolra B, Comparat D, *et al.* Photoassociative spectroscopy of Cs_2: formation of cold molecules and determination of the Cs-Cs scattering lengths. In: Aspect A, Inguscio M, Martelluci S, Chester AN, Eds. Bose-Einstein condensates and atoms lasers. Kluwer Academic/Plenum Publishers 2000
[120] Drag C, Laburthe-Tolra B, Dulieu O, *et al.* Experimental versus theoretical rates for photoassociation and for formation of ultracold molecules. IEEE J Quantum Electron 2000; 36(12): 1378-88.
[121] Pichler M, Chen H, Stwalley WC. Photoassociation spectroscopy of ultracold Cs below the $6p_{3/2}$ limit. J Chem Phys 2004; 121(14): 6779-84.
[122] Bouloufa N, Crubellier A, Dulieu O. Reexamination of the 0g- pure long-range state of Cs_2: prediction of missing levels in the photoassociation spectrum. Phys Rev A 2007; 75(5): 052501.
[123] Wester R, Kraft SD, Mudrich M, *et al.* Photoassociation inside an optical dipole trap: absolute rate coefficients and Franck Condon factors. Applied Physics B: Lasers Opt 2004; 79(8): 993-9.
[124] Ma J, Wang L, Zhao Y, Xiao L, Jia, S. High sensitive photoassociation spectroscopy of the Cs molecular 0_u^+ and 1_g long-range states below the $6s_{1/2}$ +$6p_{3/2}$ limit. J Mol Spectrosc 2009; 255(2): 106-10.
[125] Dion CM, Dulieu O, Comparat D, *et al.* Photoionization and detection of ultracold Cs_2 molecules through diffuse bands. Eur Phys J D 2002; 18(3): 365-70.
[126] Bouloufa N, Favilla E, Viteau M, *et al.* Photoionisation spectroscopy of excited states of cold caesium dimers. Mol Phys 2010. (iFirst): Available online 03 August 2010.
[127] Pillet P, Viteau M, Chotia A, *et al.* Laser cooling of molecules. World Scientific 2008.
[128] Stwalley WC, Uang Y-H, Pichler G. Pure long-range molecules. Phys Rev Lett 1978; 41(17): 1164-7.
[129] Ketterle W, Davis KB, Joffe MA, Martin A, Pritchard DE. High densities of cold atoms in a dark spontaneous-force optical trap. Phys Rev Lett 1993; 70(11): 2253-6.
[130] Bohn JL, Julienne PS. Semianalytic treatment of two-color photoassociation spectroscopy and control of cold atoms. Phys Rev A 1996 54(6): R4637-40
[131] Viteau M, Chotia A, Allegrini M, *et al.* Efficient formation of deeply bound ultracold molecules probed by broadband detection. Phys Rev A 2009; 79(2): 021402.
[132] Deiglmayr J, Aymar M, Dulieu O, Dowek D, Lucchese R. Private communication.
[133] Allouche A, Aubert-Frecon M, Dowek D, Lucchese R. Private communication.
[134] Vanhaecke N, de Souza Melo W, Laburthe-Tolra B, Comparat D, Pillet P. Accumulation of cold cesium molecules *via* photoassociation in a mixed atomic and molecular trap. Phys Rev Lett 2002; 89(6): 063001.
[135] Zahzam N, Vogt T, Mudrich M, Comparat D, Pillet P. Atom-molecule collisions in an optically trapped gas. Phys Rev Lett 2006; 96(2): 023202.
[136] Grimm R, Weidemuller M, Ovchinnikov YB. Optical dipole traps for neutral atoms. Adv At Mol, Opt Phys 2000: 42: 95-170.
[137] Takekoshi T, Patterson B M, Knize RJ. Observation of optically trapped cold cesium molecules. Phys Rev Lett 1998; 81(23): 5105.
[138] Staanum P, Kraft SD, Lange J, Wester R, Weidemuller M. Experimental investigation of ultracold atom-molecule collisions. Phys Rev Lett 2006; 96(2): 023201.
[139] Nikolov AN, Eyler EE, Wang XT, *et al.* Observation of ultracold ground-state potassium molecules. Phys Rev Lett 1999; 82(4): 703-6.
[140] Sage JM, Sainis S, Bergeman T, Demille D. Optical production of ultracold polar molecules. Phys Rev Lett 2005; 94(20): 203001.
[141] Danzl JG, Mark MJ, Haller E, *et al.* An ultracold high-density sample of rovibronic ground-state molecules in an optical lattice. Nature Phys 2010; 6(4): 265-70
[142] Morigi G, Pinkse PWH, Kowalewski M, de Vivie-Riedle R. Cavity cooling of internal molecular motion. Phys Rev Lett 2007; 99(7): 073001.
[143] Tannor DJ, Bartana A. On the interplay of control fields and spontaneous emission in laser cooling. J Phys Chem A 1999; 103(49): 10359-63.
[144] Bartana A, Kosloff R, Tannor DJ. Laser cooling of internal degrees of freedom. J Chem Phys 1997; 106(4): 1435-48.
[145] Schirmer SG. Laser cooling of internal molecular degrees of freedom for vibrationally hot molecules. Phys Rev A 2001; 63(1): 013407.

[146] Viteau M, Chotia A, Allegrini M, *et al.* Optical pumping and vibrational cooling of molecules. Science 2008; 321(5886): 232-4.

[147] Weickenmeier W, Diemer U, Wahl M, Raab M, Demtroder W, Muller W. Accurate ground state potential of Cs_2 up to the dissociation limit. J Chem Phys 1985; 82(12): 5354-63.

[148] Diemer U, Duchowicz R, Ertel M, Mehdizadeh E, Demtroder W. Doppler-free polarization spectroscopy of the $B^{-1}\Pi_u$ state of Cs_2. Chem Phys Lett 1989; 164(4): 419-26.

[149] Hertzberg, G. Atomic spectra and atomic structure. New York: Dover Publications, 1944. pp. 208.

[150] Felinto D, Bosco CAC, Acioli LH, Vianna SS. Coherent accumulation in two-level atoms excited by a train of ultrashort pulses. Opt Commun 2003; 215(1-3): 69-73.

[151] Sofikitis D, Weber S, Fioretti A, *et al.* Pulse shaping and optical pumping of the molecular vibration. New J Phys 2009; 11(5): 055037.

[152] Sofikitis D, Fioretti A, Weber S, *et al.* Broadband vibrational cooling of cold cesium molecules: theory and experiments. Chin J Chem Phys 2009; 22(1): 149-56.

[153] Sofikitis D, Horchani R, Li X, *et al.* Vibrational cooling of cesium molecules using noncoherent broadband light. Phys Rev A 2009; 80(5): 051401.

[154] Fioretti A, Sofikitis D, Horchani R, *et al.* Cold cesium molecules: from formation to cooling. Journal of Modern Optics 2009; 56(18): 2089 -99.

[155] Sofikitis D, Fioretti A, Weber S, *et al.* Vibrational cooling of cold molecules with optimised shaped pulses. Mol Phys 2010; 108(6): 795 - 810.

[156] Rieger T, Junglen T, Rangwala SA, Pinkse PW, Rempe G. Continuous loading of an electrostatic trap for polar molecules. Phys Rev Lett 2005; 95(17): 173002.

Subject Index

Gennadi Ogurtsov and Danielle Dowek (Eds)

www.ingramcontent.com/pod-product-compliance
Lightning Source LLC
Chambersburg PA
CBHW041717210326
41598CB00007B/683
* 9 7 8 1 6 0 8 0 5 6 2 1 7 *